9
Modern Mathematics for Schools

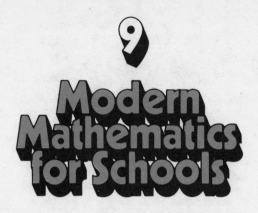

Modern Mathematics for Schools

Second Edition

Scottish Mathematics Group

Blackie

Chambers

Blackie & Son Limited
Bishopbriggs · Glasgow G64 2NZ
7 Leicester Place
London WC2H 7BP

W & R Chambers Limited
43–45 Annandale Street · Edinburgh EH7 4AZ

Designed by James W. Murray

International Standard Book Numbers
Pupils' Book
Blackie 0 216 89432 8
Chambers 0 550 75920 4
Teachers' Book
Blackie 0 216 89433 6
Chambers 0 550 75930 1

Printed in Great Britain by
Thomson Litho Ltd, East Kilbride, Scotland
Set in 10pt Monotype Times Roman

Scottish Mathematics Group

Members associated with this book

W. T. Blackburn
Dundee College of Education

W. Brodie
Trinity Academy

C. Clark
Formerly of Lenzie Academy

D. Donald
Formerly of Robert Gordon's College

R. A. Finlayson
Jordanhill College School

Elizabeth K. Henderson
Westbourne School for Girls

J. L. Hodge
Madras College

J. Hunter
University of Glasgow

R. McKendrick
Langside College

W. More
Formerly of High School of Dundee

Helen C. Murdoch
Hutchesons' Girls' Grammar School

A. G. Robertson
John Neilson High School

A. G. Sillitto
Formerly of Jordanhill College of Education

A. A. Sturrock
Grove Academy

Rev. J. Taylor
St. Aloysius' College

E. B. C. Thornton
Bishop Otter College

J. A. Walker
Dollar Academy

P. Whyte
Hutchesons' Boys' Grammar School

Preface

Book 1 of the original series *Modern Mathematics for Schools* was first published in July 1965. This revised series has been produced in order to take advantage of the experience gained in the classroom with the original textbooks and to reflect the changing mathematical needs in recent years, particularly as a result of the general move towards some form of comprehensive education.

Throughout the whole series, the text and exercises have been cut or augmented wherever this was considered to be necessary, and nearly every chapter has been completely rewritten. In order to cater more adequately for the wider range of pupils now taking certificate-oriented courses, the pace has been slowed down in the earlier books in particular, and parallel sets of A and B exercises have been introduced where appropriate. The A sets are easier than the B sets, and provide straightforward but comprehensive practice; the B sets have been designed for the more able pupils, and may be taken in addition to, or instead of, the A sets. Often from Book 4 onwards a basic exercise, which should be taken by all pupils, is followed by a harder one on the same work in order to give abler pupils an extra challenge, or further practice; in such a case the numbering is, for example, Exercise 2 followed by Exercise 2B. It is hoped that this arrangement, along with the *Graph Workbook for Modern Mathematics*, will allow considerable flexibility of use, so that while all the pupils in a class may be studying the same topic, each pupil may be working examples which are appropriate to his or her aptitude and ability.

In 1974 the first of a series of pads of expendable *Mathsheets* was published in order to provide simpler and more practical material;

each pad contains a worksheet for every exercise in the corresponding textbook. This increases still further the flexibility of the *Modern Mathematics for Schools* 'package' by providing three closely related levels of work—mathsheets, A exercises and B exercises.

Each chapter is backed up by a summary, and by revision exercises; in addition, cumulative revision exercises have been introduced at the end of alternate books. A new feature is the series of Computer Topics from Book 4 to Book 7. These form an elementary introduction to computer studies, and are primarily intended to give pupils some appreciation of the applications and influence of computers in modern society.

Books 1 to 7 provide a suitable course for many modern Ordinary Level and Ordinary Grade syllabuses in mathematics, including the University of London GCE Syllabus C, the Associated Examining Board Syllabus C, the Cambridge Local Syndicate Syllabus C, and the Scottish Certificate of Education. Books 8 and 9 complete the work for the Scottish Higher Grade Syllabus, and provide a good preparation for all Advanced Level and Sixth Year Syllabuses, both new and traditional.

Related to this revised series of textbooks are the *Modern Mathematics Newsletters* (No. 5, February 1975), the *Teacher's Editions* of the textbooks, the *Graph Workbook for Modern Mathematics*, the *Three-Figure Tables for Modern Mathematics*, and the booklets of *Progress Papers for Modern Mathematics*. These Progress Papers consist of short, quickly marked objective tests closely connected with the textbooks. There is one booklet for each textbook, containing A and B tests on each chapter, so that teachers can readily assess their pupils' attainments, and pupils can be encouraged in their progress through the course.

The separate headings of Algebra, Geometry, Arithmetic, and later Trigonometry and Calculus, have been retained in order to allow teachers to develop the course in the way they consider best. Throughout, however, ideas, material and method are integrated *within* each branch of mathematics and *across* the branches; the opportunity to do this is indeed one of the more obvious reasons for teaching this kind of mathematics in the schools—for it is *mathematics* as a whole that is presented.

Pupils are encouraged to find out facts and discover results for themselves, to observe and study the themes and patterns that pervade mathematics today. As a course based on this series of books progresses, a certain amount of equipment will be helpful, particularly in the development of geometry. The use of calculating machines, slide rules, and computers is advocated where appropriate, but these instruments are not an essential feature of the work.

While fundamental principles are emphasised, and reasonable attention is paid to the matter of structure, the width of the course should be sufficient to provide a useful experience of mathematics for those pupils who do not pursue the study of the subject beyond school level. An effort has been made throughout to arouse the interest of all pupils and at the same time to keep in mind the needs of the future mathematician.

The introduction of mathematics in the Primary School and recent changes in courses at Colleges and Universities have been taken into account. In addition, the aims, methods, and writing of these books have been influenced by national and international discussions about the purpose and content of courses in mathematics, held under the auspices of the Organisation for Economic Co-operation and Development and other organisations.

The authors wish to express their gratitude to the many teachers who have offered suggestions and criticisms concerning the original series of textbooks; they are confident that as a result of these contacts the new series will be more useful than it would otherwise have been.

Algebra

Geometry

Trigonometry

Calculus

Notation

Sets of numbers

Different countries and different authors
give different notations and definitions
for the various sets of numbers.
In this series the following are used:

E The universal set

$\emptyset$ The empty set

N The set of natural numbers $\{1, 2, 3, \ldots\}$

W The set of whole numbers $\{0, 1, 2, 3, \ldots\}$

Z The set of integers $\{\ldots, -2, -1, 0, 1, 2, \ldots\}$

Q The set of rational numbers

R The set of real numbers

The set of prime numbers $\{2, 3, 5, 7, 11, \ldots\}$

Algebra

Quadratic Equations and Functions

1 Solving quadratic equations

Do you recognise Figure 1, which was included in Book 5 to illustrate some of the applications of quadratic functions and their graphs (parabolas)?

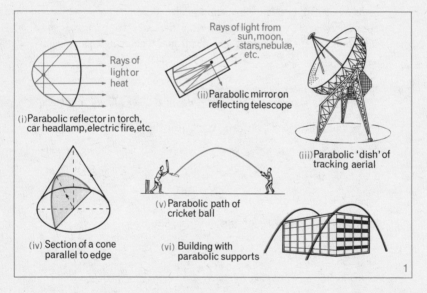

(i) Parabolic reflector in torch, car headlamp, electric fire, etc.

Rays of light or heat

Rays of light from sun, moon, stars, nebulæ, etc.

(ii) Parabolic mirror on reflecting telescope

(iii) Parabolic 'dish' of tracking aerial

(iv) Section of a cone parallel to edge

(v) Parabolic path of cricket ball

(vi) Building with parabolic supports

1

The examples in Figure 1 suggest the wide range of application of the quadratic function $f: x \rightarrow ax^2 + bx + c$ and the quadratic equation $ax^2 + bx + c = 0$ whose *solutions*, or *roots*, are the *zeros* of the function f. We assume that $a(\neq 0)$, b and c are real constants, and that $x \in R$.

In Book 6 we studied ways of solving quadratic equations. In this chapter we shall study the properties of quadratic equations and use these to investigate various problems.

Reminders. Quadratic equations can be solved by:

(i) *factorisation*

(ii) *completing the square,* in which the equation is put in the form $(x+p)^2 = q$

(iii) *the quadratic formula* for the general quadratic equation

$$ax^2 + bx + c = 0, \quad x = \frac{-b \pm \sqrt{(b^2 - 4ac)}}{2a} \text{ (proved in Book 6)}$$

(iv) *drawing the graph* of $f : x \rightarrow ax^2 + bx + c$ which is the parabola whose equation is $y = ax^2 + bx + c$.

Note. If the coefficients a, b and c are *real numbers*, the expression $ax^2 + bx + c$ is called a *real* quadratic expression and the equation a *real* quadratic equation. If a, b, c are *rational numbers*, the expression and equation are called *rational*. In this chapter, we shall be concerned only with rational quadratics, although we shall be interested in both rational and irrational linear factors.

Exercise 1

In this Exercise, x is a variable on the set of real numbers.

Solve the following quadratic equations by factorisation:

1 $x^2 - 4x + 3 = 0$ 2 $2x^2 - x - 3 = 0$ 3 $2x^2 + 3x = 0$

4 $5x - x^2 = 0$ 5 $x^2 - 4 = 0$ 6 $4x^2 - 1 = 0$

7 $x^2 + 8x + 16 = 0$ 8 $2x^2 + 3x = 35$ 9 $3x^2 = x + 10$

Solve the following by completing the square:

10 $x^2 - 2x = 3$ 11 $x^2 + 2x = 1$ 12 $x^2 - 4x - 6 = 0$

13 $2x^2 - 4x + 1 = 0$ 14 $ax^2 + bx + c = 0$, i.e. $x^2 + \dfrac{b}{a}x + \dfrac{c}{a} = 0$

Use the quadratic formula to solve the equations in questions *15–20*:

15 $x^2 - 4x + 3 = 0$ 16 $2x^2 - x - 3 = 0$ 17 $x^2 - 4 = 0$

18 $x^2 + x + 1 = 0$ 19 $2x^2 - 3x - 4 = 0$ 20 $5x^2 + 6x - 2 = 0$

21 Given $f(x) = x^2 + px + q$, show that $f(x) - f(h) = (x - h)(x + h + p)$. Deduce that if h is a root of $f(x) = 0$ (i.e. if $f(h) = 0$), then $x - h$ is a factor of $f(x)$

2 The nature of the roots of a quadratic equation: the discriminant

We know that the roots of the quadratic equation $ax^2 + bx + c = 0$ are

$$\frac{-b + \sqrt{(b^2 - 4ac)}}{2a} \quad \text{and} \quad \frac{-b - \sqrt{(b^2 - 4ac)}}{2a}.$$

(i) In *question 1* of Exercise 1, $a = 1$, $b = -4$ and $c = 3$ so $b^2 - 4ac = 16 - 12 = 4$. Hence the roots of this equation are $\frac{4 + \sqrt{4}}{2}$ and $\frac{4 - \sqrt{4}}{2}$, i.e. 3 and 1, which are *real* and *rational*.

(ii) In *question 7*, $a = 1$, $b = 8$ and $c = 16$ so $b^2 - 4ac = 64 - 64 = 0$, and there is only one root, -4. For reasons that will become apparent in Section 4 it is convenient to say that the equation has two *equal* roots -4, -4 which are *real* and *rational*.

(iii) In *question 12*, $a = 1$, $b = -4$ and $c = -6$ so $b^2 - 4ac = 16 + 24 = 40$. Hence the roots are $\frac{4 + \sqrt{40}}{2}$ and $\frac{4 - \sqrt{40}}{2}$, i.e. $2 + \sqrt{10}$ and $2 - \sqrt{10}$, which are *real* and *irrational*.

(iv) In *question 18*, $a = 1$, $b = 1$ and $c = 1$ so $b^2 - 4ac = 1 - 4 = -3$. Hence the roots are $\frac{-1 + \sqrt{(-3)}}{2}$ and $\frac{-1 - \sqrt{(-3)}}{2}$. But $\sqrt{(-3)}$ is not a real number so the roots are *not real*.

Without solving the quadratic equation $ax^2 + bx + c = 0$, it is evident that we can determine the *nature* of its roots, i.e. whether they are real or not, rational or not, equal or not, according as the number $b^2 - 4ac$ is negative or not negative. Since it discriminates between the different types of roots, $b^2 - 4ac$ is called the *discriminant* of the equation.

The discriminant $b^2 - 4ac$ indicates the nature of the roots as follows:

(i) $b^2 - 4ac \geqslant 0$ ⇔ *roots are real.* (If $b^2 - 4ac$ is a perfect square, roots are rational, otherwise irrational.)

(ii) $b^2 - 4ac = 0$ ⇔ *roots are equal.* (Also real and rational.)

(iii) $b^2 - 4ac < 0$ ⇔ *roots are not real.*

Example. Without solving the equations, find the nature of the roots of each of the following quadratic equations:

(i) $3x^2 - 5x - 2 = 0$ (ii) $9x^2 - 6x + 1 = 0$ (iii) $x^2 + 2x + 3 = 0$.

Comparing each equation with $ax^2 + bx + c = 0$,

(i) $b^2 - 4ac = (-5)^2 - 4(3)(-2) = 25 + 24 = 49$.

$b^2 - 4ac > 0$ and a perfect square, so roots are *real* and *rational*.

(ii) $b^2 - 4ac = (-6)^2 - 4(9)(1) = 36 - 36 = 0$.

$b^2 - 4ac = 0$, so roots are *equal*, *real* and *rational*.

(iii) $b^2 - 4ac = 2^2 - 4(1)(3) = 4 - 12 = -8$.

$b^2 - 4ac < 0$, so roots are *not real*.

Exercise 2

By comparing with $ax^2 + bx + c = 0$ and considering the discriminant $b^2 - 4ac$, find the nature of the roots of each of these equations:

1 $x^2 + 5x + 4 = 0$	*2* $x^2 + 5x + 7 = 0$	*3* $4x^2 + 12x + 9 = 0$
4 $x^2 - x - 1 = 0$	*5* $2x^2 + x - 3 = 0$	*6* $x^2 + x + 2 = 0$
7 $2x^2 - 5x + 3 = 0$	*8* $x^2 - 7x + 1 = 0$	*9* $3x^2 - 6x + 3 = 0$
10 $2x^2 - x + 1 = 0$	*11* $t^2 + 2t - 15 = 0$	*12* $9y^2 + 6y + 1 = 0$
13 $4c^2 - 3c + 4 = 0$	*14* $6h^2 - 5h + 1 = 0$	*15* $2x^2 + 3x + 4 = 0$
16 $10x^2 - 5x - 1 = 0$	*17* $x^2 - 18x + 81 = 0$	*18* $5x^2 + 11x + 7 = 0$

19   Copy this table, and enter 'yes' or 'no' as appropriate:

Discriminant	Roots are:			
	real	rational	irrational	equal
$b^2 - 4ac > 0$, and perfect square				
$b^2 - 4ac = 0$				
$b^2 - 4ac < 0$				
$b^2 - 4ac > 0$, not perfect square				

Topic to explore

Find out about the 'number' i whose square is -1.

---

3   Using the discriminant

Example. Find p, given that the quadratic equation
$px^2 + (p+8)x + 9 = 0$ has: (i) equal roots (ii) real and distinct roots.

Discriminant $= `b^2 - 4ac` = (p+8)^2 - 36p = p^2 - 20p + 64$

(i) For equal roots,

$b^2 - 4ac = 0$
$\Leftrightarrow p^2 - 20p + 64 = 0$
$\Leftrightarrow (p-4)(p-16) = 0$
$\Leftrightarrow p = 4$ or $p = 16$

(ii) For real distinct roots,

$b^2 - 4ac > 0$
$\Leftrightarrow p^2 - 20p + 64 > 0$
$\Leftrightarrow (p-4)(p-16) > 0$
$\Leftrightarrow p < 4$ or $p > 16$

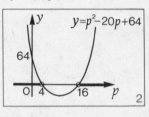

(Fig. 2)

Exercise 3

1   Find k, given that each of the following equations has equal roots:

a $x^2 - 8x + k = 0$ b $kx^2 - 12x + 9 = 0$ c $x^2 + kx + 16 = 0$

2   Find k for each equation in question *1* if the roots are real.

3   Find m, given that the equation $x^2 + 2mx + 9 = 0$ has:

a   equal roots   b   no real roots

4   Find p, given that $x^2 + (p+1)x + 9 = 0$ has equal roots.

5   Find q, given that $qx^2 + 2x + q = 0$ has real, distinct roots.

6   Find p, given that $(p+1)x^2 - 2(p+3)x + 3p = 0$ has equal roots.

7   Find a, given that $a^2x^2 + 2(a+1)x + 4 = 0$ has equal roots. Then solve the equation for each case.

8   Find the condition for $(px+q)^2 = 8x$ to have equal roots.

9   Find the condition for $x^2 + (x+c)^2 = 8$ to have equal roots.

10   Show that if c is real, the roots of $(x-1)(x-2) = c^2$ are real.

11   Show that the roots of $k(x+1)(x+4) = x$ are not real if $\frac{1}{9} < k < 1$.

12a   Find m if $(2m-1)x^2 + (m+1)x + 1 = 0$ has equal roots.

   b   Find the nature of the roots if m lies between these values.

13   Find m, given that $x^2 + (mx-5)^2 = 9$ has equal roots.

14a   If $\dfrac{x^2 - 3x + 3}{x-2} = t$, show that $x^2 - (t+3)x + (2t+3) = 0$.

   b   Given that x is real, prove that $t \leqslant -1$ or $t \geqslant 3$.

15   Given that $\dfrac{x^2 + 4x + 10}{2x+5} = n$, form a quadratic equation in x, and hence show that $n \leqslant -3$ or $n \geqslant 2$ for real x.

16   If $y = \dfrac{x^2 + x + 1}{x^2 - x + 1}$, show that $(1-y)x^2 + (1+y)x + (1-y) = 0$.

   Hence show that if x is real, the maximum and minimum values of y are 3 and $\frac{1}{3}$ respectively.

17   Given that $kx^2 - 10x + 1 = 0$ has rational roots, and that k is a positive integer, find all the possible replacements for k.

---

4   Tangents to curves

Reminders. The equations of a straight line include:

(i) $y = mx + c$, where the line has gradient m and passes through the point $(0, c)$.

(ii) $y - b = m(x-a)$, where the line has gradient m and passes through the point (a, b).

   Example 1. Find the equation of the tangent to the parabola $y = x^2$ which has gradient 1.

   Figure 3 shows the parabola $y = x^2$ and a set of lines with gradient 1. Each line of this set has an equation of the form $y = x + c$, where $c \in R$. Such a line meets the parabola where $y = x^2$ *and* $y = x + c$, i.e. $x^2 - x - c = 0$ *and* $y = x + c$.

From Figure 3, it can be seen that the intersections will be two distinct points, two coincident points, or no points at all. These occur when the discriminant of the equation $x^2 - x - c = 0$ is positive, zero, or negative, respectively.

Evidently the case of 'two coincident points' is a limiting case where the line touches the curve, i.e. is a tangent to the curve.

For tangency, therefore, we must choose c so that the equation $x^2 - x - c = 0$ has equal roots, i.e. the discriminant is zero.

$$`b^2 - 4ac = 0' \quad \Leftrightarrow \quad 1 + 4c = 0 \quad \Leftrightarrow \quad c = -\tfrac{1}{4}.$$

Hence the equation of the tangent of gradient 1 is $y = x - \tfrac{1}{4}$.

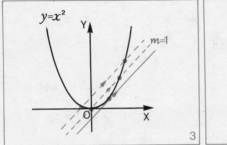 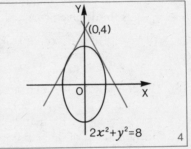

Example 2. Find the equations of the tangents from the point $(0, 4)$ to the ellipse $2x^2 + y^2 = 8$ (see Figure 4).

The line through $(0, 4)$ with gradient m has equation

$$y - 4 = m(x - 0)$$
$$\Leftrightarrow \quad y = mx + 4.$$

This line cuts the ellipse where

$$2x^2 + (mx + 4)^2 = 8$$
$$2x^2 + m^2x^2 + 8mx + 16 = 8$$
$$(m^2 + 2)x^2 + 8mx + 8 = 0.$$

For the line to be a tangent, the discriminant of this quadratic equation in x must be zero, i.e. $b^2 - 4ac = 0$.

Hence
$$(8m)^2 - 4(8)(m^2 + 2) = 0$$
$$\Leftrightarrow \quad 64m^2 - 32m^2 - 64 = 0$$
$$\Leftrightarrow \quad 32m^2 = 64$$
$$\Leftrightarrow \quad m = \pm\sqrt{2}.$$

So the equations of the tangents are $y = \sqrt{2}x + 4$ and $y = -\sqrt{2}x + 4$.

Exercise 4

In questions *1–6*, find the equations of the tangents to the given curves, with gradients as shown. Illustrate each with a sketch.

1 $y = x^2, m = 2$ **2** $y = 2x^2, m = -8$ **3** $xy = 1, m = -4$

4 $y = x^2 + 1, m = 2$ **5** $y^2 = 4x, m = 1$ **6** $x^2 + y^2 = 8, m = -1$

In questions *7–12*, find the equations of the tangents from the given points to the given curves. Illustrate each with a sketch.

7 $(0, -4), y = x^2$ **8** $(0, -2), y = 8x^2$ **9** $(0, 0), y = x^2 + 4$

10 $(1, 1), xy = 1$ **11** $(0, -4), x^2 + y^2 = 4$ **12** $(0, 5), x^2 + y^2 = 9$

13   By considering the intersections of the line $y = k$, where k is a real number, with the parabola $y = x^2 + 4x + 5$, find k such that $y = k$ is a tangent to the parabola. Hence find the coordinates of the turning point of the parabola.

14   Repeat question *13* for each of the following parabolas:

a $y = x^2 - 6x + 8$ *b* $y = 8x - 7 - x^2$

15   Find c such that the line $y = x + c$ is a tangent to the curve $y = x^2 - 3x$.
         What are the coordinates of the point of contact?

16   Prove that if the line $y = x + k$ is a tangent to the ellipse $x^2 + 9y^2 = 9$, then $k = \pm\sqrt{10}$.

---

5   *Sketching the graph of a quadratic function*

The graph of every quadratic function f defined by $f(x) = ax^2 + bx + c$ is a parabola $y = ax^2 + bx + c$ situated relative to the x-axis in one of the six ways shown in Figure 5

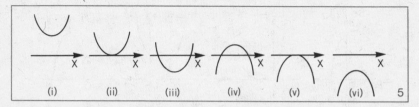

The *axis* of the parabola is parallel to, or coincident with, the y-axis. The graph cuts, or touches, the x-axis where $ax^2 + bx + c = 0$. Since the discriminant of this quadratic equation reveals the nature of the roots of the equation, it also provides information about the points of intersection of the graph and the x-axis.

$b^2 - 4ac > 0 \Leftrightarrow$ two distinct points of intersection ((iii) and (iv).)
$b^2 - 4ac = 0 \Leftrightarrow$ two coincident points of intersection ((ii) and (v).)
$b^2 - 4ac < 0 \Leftrightarrow$ no points of intersection ((i) and (vi).)

Knowing that the graph of f is a parabola, we can sketch the curve $y = ax^2 + bx + c$ by means of the following techniques which are now at our disposal:

(i) $ax^2 + bx + c$ factorises

1. Find the point where the curve cuts the y-axis by putting $x = 0$.
2. Find the points where the curve cuts the x-axis by putting $y = 0$.
3. Find the turning point from considerations of symmetry.

(ii) $ax^2 + bx + c$ *does not factorise*

1. Find the point where the curve cuts the y-axis.
2. Find the turning point by calculus.

Note also that for large x, $y \doteqdot ax^2$, so for:
a positive, y is large and positive as in Figure 5 (i), (ii) and (iii);
a negative, y is large and negative as in Figure 5 (iv), (v) and (vi).

Example. Sketch curves (i) to (iv) below.

(i) $y = 8 - 2x - x^2$

If $x = 0$, $y = 8$.
If $y = 0$, $8 - 2x - x^2 = 0$
$\Leftrightarrow (4 + x)(2 - x) = 0$
$\Leftrightarrow x = -4$ or 2.
Turning point is $(-1, 9)$.

(ii) $y = 4x^2 - 4x + 1$

If $x = 0$, $y = 1$.
If $y = 0$, $4x^2 - 4x + 1 = 0$
$\Leftrightarrow (2x - 1)(2x - 1) = 0$
$\Leftrightarrow x = \frac{1}{2}$ or $\frac{1}{2}$.
Turning point is $(\frac{1}{2}, 0)$.

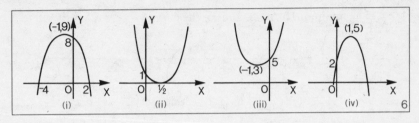

(iii) $y = 2x^2 + 4x + 5$

    If $x = 0$, $y = 5$.
    If $y = 0$, $2x^2 + 4x + 5 = 0$.
    $b^2 - 4ac = 16 - 40 = -24$,
    so curve does not cut x-axis.

$$\frac{dy}{dx} = 4x + 4.$$

For turning point, $\dfrac{dy}{dx} = 0$

$\Rightarrow \quad x = -1$, giving $(-1, 3)$.

(iv) $y = 2 + 6x - 3x^2$

    If $x = 0$, $y = 2$.
    If $y = 0$, $2 + 6x - 3x^2 = 0$.
    $b^2 - 4ac = 36 + 24 = 60$,
    so roots are real, irrational.

$$\frac{dy}{dx} = 6 - 6x.$$

For turning point, $\dfrac{dy}{dx} = 0$

$\Rightarrow \quad x = 1$, giving $(1, 5)$.

Exercise 5

1   Which parabolas in Figure 5 correspond to the following equations?

  a $y = x^2 + 4x + 4$ *b* $y = 9 - x^2$ *c* $y = 2x^2 + x + 3$

  d $y = (1 - x)^2$ *e* $y = x^2 - x - 2$ *f* $y = -x^2 + 2x - 5$

    Sketch the parabolas in questions **2–16**.

2 $y = x^2 - 4x + 3$ **3** $y = 8 - 2x - x^2$ **4** $y = x^2 + 2x + 1$

5 $y = 4x - x^2$ **6** $y = x^2 + 6x$ **7** $y = -x^2 + 6x - 9$

8 $y = x^2 - 2x + 5$ **9** $y = 1 - 2x - x^2$ **10** $y = x^2 + 2x - 3$

11 $y = 6 + x - x^2$ **12** $y = (2x - 3)^2$ **13** $y = 3x^2 + 6x + 5$

14 $y = -x^2$ **15** $y = (1 + x)(3 - x)$ **16** $y = (2x - 9)(2x + 7)$

17   Which of the following are true and which are false?

  a The curve $y = (x - 1)(x - 3)$ is symmetrical about the line $x = 2$.
  b The curve $y = 1 - x^2$ does not cut the x-axis.

c   The minimum turning point of the curve $y = x^2 - 25$ is $(0, 0)$.
d   The curve $y = 1 - 2x + x^2$ touches the x-axis.
e   The image under reflection in the x-axis of the curve $y = x^2 - 4$ is the curve $y = 4 - x^2$.

With the aid of sketch graphs solve the following inequations ($x \in R$). Indicate the solution sets in your diagrams.

18   $(x-1)(x-5) < 0$ 19 $(1+x)(3-x) \geqslant 0$ 20 $x(x-6)+8 > 0$

21   $4x^2 - 12x + 5 \geqslant 0$ 22 $12 + x - x^2 > 0$ 23 $2x^2 + 4x + 1 > 0$

---

6   The sum and product of the roots of a quadratic equation

The quadratic formula enables us to find relationships between the roots of the quadratic equation $ax^2 + bx + c = 0$ and the coefficients a, b and c.

If α and β are the roots of the quadratic equation $ax^2 + bx + c = 0$, then

$$\text{(i) } \alpha + \beta = -\frac{b}{a} \quad \text{and} \quad \text{(ii) } \alpha\beta = \frac{c}{a}.$$

Proof. Method 1.

The roots of $ax^2 + bx + c = 0$ are $x = \dfrac{-b \pm \sqrt{(b^2 - 4ac)}}{2a}$.

Let $\alpha = \dfrac{-b + \sqrt{(b^2 - 4ac)}}{2a}$ and $\beta = \dfrac{-b - \sqrt{(b^2 - 4ac)}}{2a}$.

(i) $\alpha + \beta$

$$= \frac{-b + \sqrt{(b^2 - 4ac)} - b - \sqrt{(b^2 - 4ac)}}{2a}$$

$$= \frac{-2b}{2a}$$

$$= -\frac{b}{a} \ .$$

(ii) $\alpha\beta$

$$= \frac{[-b+\sqrt{(b^2-4ac)}][-b-\sqrt{(b^2-4ac)}]}{4a^2}$$

$$= \frac{(-b)^2-(b^2-4ac)}{4a^2}, \text{ a difference of squares}$$

$$= \frac{4ac}{4a^2}$$

$$= \frac{c}{a}$$

Method 2.

If α and β are the roots of the quadratic equation $ax^2+bx+c=0$,

i.e. of $\quad x^2+\dfrac{b}{a}x+\dfrac{c}{a}=0,\quad$ then the equations

$$x^2+\frac{b}{a}x+\frac{c}{a}=0 \quad \text{and} \quad (x-\alpha)(x-\beta)=0 \text{ are identical,}$$

and so $\quad x^2+\dfrac{b}{a}x+\dfrac{c}{a}=0 \quad \text{and} \quad x^2-(\alpha+\beta)x+\alpha\beta=0 \text{ are identical.}$

Comparing coefficients of x, $\quad \alpha+\beta=-\dfrac{b}{a}$.

Comparing constant terms, $\quad \alpha\beta=\dfrac{c}{a}$.

Notice that when $ax^2+bx+c=0$ is written in the form $x^2+\dfrac{b}{a}x+\dfrac{c}{a}=0$, the sum of the roots is the negative of the coefficient of x and the product of the roots is the constant term.

Example 1. Given that α and β are the roots of $2x^2-5x+6=0$, find the values of: $\quad \boldsymbol{a} \quad \dfrac{1}{\alpha}+\dfrac{1}{\beta} \qquad \boldsymbol{b} \quad (\alpha-\beta)^2$.

We know that $\alpha+\beta=-\dfrac{b}{a}=-\dfrac{(-5)}{2}=\dfrac{5}{2}$ and $\alpha\beta=\dfrac{c}{a}=\dfrac{6}{2}=3$.

a $\dfrac{1}{\alpha}+\dfrac{1}{\beta}=\dfrac{\beta+\alpha}{\alpha\beta}=\dfrac{\frac{5}{2}}{3}=\dfrac{5}{6}$

b $(\alpha-\beta)^2=\alpha^2-2\alpha\beta+\beta^2=\alpha^2+2\alpha\beta+\beta^2-4\alpha\beta=(\alpha+\beta)^2-4\alpha\beta$
$=(\tfrac{5}{2})^2-(4\times3)=\tfrac{25}{4}-12=-5\tfrac{3}{4}$

Example 2. Find a quadratic equation with roots $\frac{1}{2}$, $-\frac{3}{2}$, giving the equation in *standard form* $ax^2+bx+c=0$.

Method 1 (Using factors)

$x=\tfrac{1}{2}$ or $x=-\tfrac{3}{2}$

$\Leftrightarrow$ $2x=1$ or $2x=-3$

$\Leftrightarrow$ $2x-1=0$ or $2x+3=0$

$\Leftrightarrow$ $(2x-1)(2x+3)=0$

$\Leftrightarrow$ $4x^2+4x-3=0$

Method 2 (Using sum and product of roots)

Let the equation be $ax^2+bx+c=0$

i.e. $x^2+\dfrac{b}{a}x+\dfrac{c}{a}=0,\quad a\neq0$

Now $-\dfrac{b}{a}=\dfrac{1}{2}+\left(-\dfrac{3}{2}\right)=-1,$

and $\dfrac{c}{a}=\left(\dfrac{1}{2}\right)\left(-\dfrac{3}{2}\right)=-\dfrac{3}{4}.$

So the required equation is

$$x^2+x-\tfrac{3}{4}=0\ \Leftrightarrow\ 4x^2+4x-3=0$$

Exercise 6

Use the formulae for $\alpha+\beta$ and $\alpha\beta$ to write down the sum and product of the roots of each of these equations:

1 $x^2-5x+6=0$ 2 $x^2+5x+6=0$ 3 $x^2+3x-1=0$

4 $x^2-2x-4=0$ 5 $3x^2-2x+1=0$ 6 $4x^2-3x-2=0$

7 $x^2+px+q=0$ 8 $lx^2-mx+n=0$ 9 $ax^2+2hx+b=0$

10 $x^2+5=0$ 11 $x^2-3x=0$ 12 $5x=1-2x^2$

13 If α and β are the roots of $x^2-2x+4=0$, find the values of:

a $\alpha+\beta$ and $\alpha\beta$ b $\dfrac{1}{\alpha}+\dfrac{1}{\beta}$ c $\alpha^2\beta+\alpha\beta^2$ d $\alpha^2+\beta^2$

14 If α and β are the roots of $3x^2+x+2=0$, find the values of:

a $\alpha+\beta$ and $\alpha\beta$ b $\dfrac{1}{\alpha}+\dfrac{1}{\beta}$ c $\alpha^2+\beta^2$ d $\dfrac{\alpha}{\beta}+\dfrac{\beta}{\alpha}$

15   If α and β are the roots of $x^2 - 2x - 1 = 0$, find the values of:

a $\alpha^2 + \beta^2$ b $(\alpha - \beta)^2$ c $\alpha^4 + \beta^4$ d $\dfrac{1}{\alpha + 1} + \dfrac{1}{\beta + 1}$

16   Given that α and β are the roots of $x^2 - px + q = 0$, express the following in terms of p and q:

a $\alpha + \beta$ and $\alpha\beta$ b $\alpha^2 - \alpha\beta + \beta^2$ c $\dfrac{\alpha + \beta}{\alpha\beta}$ d $\dfrac{1}{\alpha^2} + \dfrac{1}{\beta^2}$

17   Given that α and β are the roots of $ax^2 + bx + c = 0$, show that $(\alpha - \beta)^2 = \dfrac{1}{a^2} \times$ discriminant.

Exercise 7B

1   Find m if one root of $x^2 + 6x + m = 0$ is double the other. (Let the roots be α and 2α.)

2   Find m if one root of $mx^2 - 16x + 3 = 0$ is three times the other, and hence find the roots.

3   In the equation $4x^2 - 16x + m = 0$, one root is 3 greater than the other root. Find m and the two roots. (Let the roots be α and $\alpha + 3$.)

4   Using the factor method, find a quadratic equation in its simplest form with roots:

a $3, 4$ b $2, -1$ c $-2, -3$ d $\frac{1}{2}, -1$ e $\frac{4}{3}, \frac{1}{3}$

5   Using the sum and product method, find a quadratic equation in standard form with roots:

a $2, 5$ b $-2, 7$ c $1 \pm \sqrt{2}$ d $5 \pm \sqrt{3}$

6   The sum of the roots of a quadratic equation is $\frac{7}{6}$ and the product of the roots is $-\frac{1}{2}$. Find the roots.

7 ·   Given that α and β are the roots of $3x^2 + 8x - 4 = 0$, find an equation whose roots are $\dfrac{1}{\alpha}$ and $\dfrac{1}{\beta}$.

8   Given that α, β are the roots of $3x^2 - 4x - 2 = 0$, find in standard form an equation whose roots are $\alpha - 2, \beta - 2$.

9   If α, β are the roots of $4x^2 - 2x + 3 = 0$, find in simplest form an equation whose roots are $\alpha + 1$, $\beta + 1$.

10   If α, β are the roots of $2x^2 - x - 5 = 0$, find in standard form an equation whose roots are α^2, β^2.

11   A function f is defined by $f(x) = 3x^2 - 12x + 6$. If the graph of f cuts the x-axis at $(p, 0)$ and $(q, 0)$, which of the following are true?

   a $pq = 2$ b $p + q = 4$ c p and q are surds   d $p - q = \pm 2\sqrt{2}$

12   One root of $x^2 + px + 48 = 0$ is 6 and the equation $x^2 + px + q = 0$ has equal roots. Find q.

As in Exercise 1, question *21*, it can be shown that:

α is a root of $ax^2 + bx + c = 0$ $\Leftrightarrow$ $x - \alpha$ is a factor of $ax^2 + bx + c$.

Hence, knowing the roots of a quadratic equation, the factors of any quadratic expression of the form $ax^2 + bx + c$ can be found even if the factors are not rational. The technique is also useful when a trinomial is difficult to factorise by any former method.

Example. Factorise $4x^2 - 8x + 1$.

Consider the equation $4x^2 - 8x + 1 = 0$.

Using the quadratic formula, $x = \dfrac{8 \pm \sqrt{(64 - 16)}}{8} = \dfrac{2 \pm \sqrt{3}}{2}$.

Hence

$$4x^2 - 8x + 1 = 4\left(x - \frac{2 + \sqrt{3}}{2}\right)\left(x - \frac{2 - \sqrt{3}}{2}\right)$$

$$= 4\left(x - 1 - \frac{1}{2}\sqrt{3}\right)\left(x - 1 + \frac{1}{2}\sqrt{3}\right)$$

Exercise 8B

Factorise the following expressions by first solving the corresponding quadratic equations:

1   $x^2 - 100$ 2 $x^2 - 2$ 3 $x^2 + 2x - 1$

4   $x^2 + x - 1$ 5 $4x^2 - 4x - 1$ 6 $2x^2 + 2x - 11$

7   $2x^2 + 7x - 15$ 8 $3x^2 + 6x + 1$ 9 $7x^2 - 4x - 4$

Summary

1   *Quadratic equations* can be solved by:

    (i) factorisation

    (ii) completing the square, by expressing the equation as $(x+p)^2 = q$

    (iii) using the quadratic formula. For the quadratic equation $ax^2 + bx + c = 0$,

$$x = \frac{-b \pm \sqrt{(b^2 - 4ac)}}{2a}$$

    (iv) using the graph of $y = ax^2 + bx + c$.

2   *The nature of the roots* of quadratic equation $ax^2 + bx + c = 0$ is indicated by the *discriminant* $b^2 - 4ac$:

    (i) $b^2 - 4ac > 0$ ⇔ roots are *real*. (If $b^2 - 4ac$ is a perfect square, roots are rational, otherwise irrational.)

    (ii) $b^2 - 4ac = 0$ ⇔ roots are *equal* (also real and rational).

    (iii) $b^2 - 4ac < 0$ ⇔ roots are *not real*.

3   *Tangents to curves*

A tangent to a curve meets the curve at *two coincidental points*. So the quadratic equation formed from the system of equations of the curve and line has *equal roots*, for which $b^2 - 4ac = 0$.

4   *Sketching the graph of a quadratic function*

The discriminant of the associated quadratic equation indicates the number of points of intersection of the graph and the x-axis. Then sketch the graph by finding:

    (i) its points of intersection with the x- and y-axes

    (ii) its turning point, using symmetry or calculus.

5   *The sum and product of the roots of a quadratic equation*

For $ax^2 + bx + c = 0$, $\alpha + \beta = -\dfrac{b}{a}$ and $\alpha\beta = \dfrac{c}{a}$.

6   *Formation of a quadratic equation, given the roots α and β*

Using factors, or sum and product of roots, the required equation is $x^2 - (\alpha + \beta)x + \alpha\beta = 0$.

Polynomials, the Remainder Theorem and Applications

1 Polynomials; the value of a polynomial

$x^2 + 5x - 2$ and $2x^5 - 6x^3 + 11x$ are called *polynomials* in x of degree two and five respectively; the *degree* is the index of the highest power of x in the polynomial. If $a_n, a_{n-1}, \ldots, a_0$ are constants,

$$a_n x^n + a_{n-1} x^{n-1} + \ldots + a_1 x + a_0$$

is a polynomial in x of degree n, provided that $n \in W$ and $a_n \neq 0$.

Notice that in a polynomial all the indices are greater than or equal to zero. The number a_k is called the *coefficient* of the term in x^k and a_0 is the *constant term*. In the form shown, the polynomial is arranged in descending powers of x.

A polynomial is determined by its coefficients. For example, in $8x^3 + 5x - 2$ the coefficient of x^3 is 8, of x^2 is 0, of x is 5, and the constant term is -2.

An expression such as $(1-x)(2+x+x^2) + 3x + 7$ is also called a polynomial since it can be written $-x^3 + 2x + 9$. Denoting the polynomial by $f(x)$, read 'f of x', the *value* of the polynomial when x is replaced by -1 is $f(-1)$.

$$f(-1) = -(-1)^3 + 2(-1) + 9 = 1 - 2 + 9 = 8.$$

Exercise 1

1 Give the degree of each of the following polynomials:

  a $4x^2 + 3x + 2$ b $x^3 - x - 11$ c $5 - x^2 + 3x^4$ d $8x + 3$ e 4

2 Arrange the following in descending powers of x, and state their degrees:

  a $x^3 + 1 + x^2 + x$ b $4x + x^4 + 1$ c $x(1-x)(1+x)$

  d $(5x-1)^2$ e $(2x-3)(2-3x)$ f $(x+1)(x+2)(x+3)$

20

3 Find the coefficient of:

    a x^2 in $(3x+4)(1-2x)$ *b* x in $(x+2)(2x-1)$

    c x^3 in $(2x-9)(3x^2+11)$ *d* x in $(x+1)(x^2+x+5)$

4 Write down the other factor in each of the following:

    a $2x^2-5x-7 = (x+1)(.....)$ *b* $4x^2-20x+9 = (2x-1)(.....)$

    Calculate:

5 $f(2)$ when $f(x) = x^3-5x-1$ **6** $f(4)$ when $f(x) = 2x^2-x-8$

7 $f(-1)$ when $f(x) = x^4-x^2-1$ **8** $f(-3)$ when $f(x) = x^3-8x+3$

Calculate the values of the following polynomials for the given replacements for x:

9 x^3+7x^2-4x+3; 2   **10** $5x^4+7x^2+3x+1$; -1

11 $4x^2-6x+2$; $\frac{1}{2}$ **12** x^3-x+1; $-\frac{1}{3}$

13 x^3+7x^2-4x+3; 5   **14** $7x^4-20x^2+15x+2$; -2

---

2   *Another method of calculating the value of a polynomial*

The substitution method in Exercise 1 is a tedious way of calculating the value of a polynomial, except in simple cases. In this section we find an alternative method of evaluation which can be applied to all polynomials.

Let $f(x) = ax^3+bx^2+cx+d$ and suppose that we require $f(h)$, where h is any real number. The substitution method requires the calculation of $f(h) = ah^3+bh^2+ch+d$.

Now ah^3+bh^2+ch+d can be expressed in *nested form* as follows:

$$ah^3+bh^2+ch+d$$
$$= (ah^2+bh+c)h+d$$
$$= [(ah+b)h+c]h+d$$

Reversing the process, we can *form* $ah^3 + bh^2 + ch + d$ in this way:

(i) *multiply a by h and add b* ....$ah + b$.

(ii) *multiply ah + b by h and add c* ....$ah^2 + bh + c$.

(iii) *multiply $ah^2 + bh + c$ by h and add d* ....$ah^3 + bh^2 + ch + d$.

The working for this 'multiply and add' procedure can be set down as in the following scheme:

$$
\begin{array}{c|cccc}
h & a & b & c & d \\
 & & \nearrow ah & \nearrow ah^2 + bh & \nearrow ah^3 + bh^2 + ch \\
 & & \times h & \times h & \times h \\
\hline
 & a & ah + b & ah^2 + bh + c & ah^3 + bh^2 + ch + d
\end{array}
$$

Example. Calculate $f(3)$ when $f(x) = 2x^3 + 4x^2 - 18$.

$$
\begin{array}{c|cccc}
3 & 2 & 4 & 0 & -18 \\
 & & \nearrow 6 & \nearrow 30 & \nearrow 90 \\
 & & \times 3 & \times 3 & \times 3 \\
\hline
 & 2 & 10 & 30 & 72
\end{array}
$$

, so $f(3) = 72$.

Note. (i) The first row to the right of the vertical line contains the coefficient of each power of x (in descending order). If any power is 'missing', i.e. its coefficient is zero, then a zero must be inserted.

(ii) Each arrow indicates a multiplication by 3, which is then followed by an addition.

(iii) The method is particularly suitable for use with a slide rule or desk calculator.

Check. $f(3) = 2(3)^3 + 4(3)^2 - 18 = 54 + 36 - 18 = 72$.

Exercise 2

Use the above scheme to calculate:

1 $f(3)$ when $f(x) = 2x^2 + 4x + 6$ 2 $f(4)$ when $f(x) = x^2 + 2x - 5$

3 $f(-1)$ when $f(x) = 3x^2 + 7x - 2$ 4 $f(5)$ when $f(x) = x^2 - 3x + 4$

Calculate the values of the following polynomials for the given replacements for x:

5 $x^3 + 2x^2 + 3x - 4$; 5   6 $2x^3 + 4x^2 + 6x + 8$; 3

7    $x^4 + 3x^3 - x^2 + 7x + 25; -4$ 8 $2x^3 + 3x^2 + 5; -2$ (note $0x$)

9    $x^3 - x + 1; 4$ 10 $2x^3 - 3x^2 + 9x + 12; \frac{1}{2}$

11   $5x^4 + 2x^3 - 4x^2 + 1; 0 \cdot 6$ 12 $5x^3 + 4x^2 + 3 \cdot 68; -0 \cdot 4$

---

3   Division of polynomials

Example 1. Calculate $3693 \div 15$ by long division.

$$\begin{array}{r} 246 \\ 15)\overline{3693} \\ 3000 \\ \hline 693 \\ 600 \\ \hline 93 \\ 90 \\ \hline 3 \end{array}$$

This shows:

$3693 = (15 \times 200) + 693$

$\qquad = (15 \times 200) + (15 \times 40) + 93$

$\qquad = (15 \times 200) + (15 \times 40) + (15 \times 6) + 3*$

$\qquad = (15 \times 246) + 3.$

* *The division ends here since 3 is less than 15.*

Hence $3693 = (15 \times 246) + 3.$

We call 15 the *divisor*, 246 the *quotient* and 3 the *remainder*.

Example 2. Divide $2x^2 + 3x - 4$ by $x - 2$.

$$\begin{array}{r} 2x + 7 \\ x - 2)\overline{2x^2 + 3x - 4} \\ 2x^2 - 4x \\ \hline 7x - 4 \\ 7x - 14 \\ \hline 10 \end{array}$$

This shows:

$2x^2 + 3x - 4 = (x - 2)2x + 7x - 4$

$\qquad = (x - 2)2x + (x - 2)7 + 10*$

$\qquad = (x - 2)(2x + 7) + 10$

* *The division ends here since 10 is of lower degree than $x - 2$.*

Hence $2x^2 + 3x - 4 = (x - 2)(2x + 7) + 10.$

We call $x - 2$ the *divisor*, $2x + 7$ the *quotient* and 10 the *remainder*.

In this chapter we shall be concerned mainly with the division of polynomials by polynomials of the first degree, i.e. *linear polynomials* such as $x-2$, $x+4$, $x-\frac{1}{2}$ and so on. In other words, the divisors will be of the form $x-h$, where h is a real number. Our immediate aim is to find quick and easy ways of computing quotients and remainders.

Exercise 3

In questions *1–6* do the divisions and present your results in the form (divisor × quotient) + remainder, as in Example 1, i.e. $3693 = (15 \times 246) + 3$.

1 $36 \div 7$ *2* $100 \div 12$ *3* $136 \div 5$

4 $867 \div 9$ *5* $543 \div 13$ *6* $2046 \div 31$

In questions *7–14* do the divisions and present your results in the final form obtained in Example 2, i.e.

$$2x^2 + 3x - 4 = (x-2)(2x+7) + 10.$$

7 $(6x+8) \div (x-3)$ *8* $(2x-1) \div (x+4)$

9 $(x^2 + 5x + 4) \div (x+2)$ *10* $(8x^2 - 4x + 11) \div (x+5)$

11 $(6x^2 - 28x - 15) \div (x-5)$ *12* $(x^3 + 2x^2 + 3x + 6) \div (x-2)$

13 $(x^3 + 6x^2 + 3x - 15) \div (x+3)$ *14* $(2x^3 - 4x^2 - 5x + 9) \div (x+1)$

15   Find the remainder on dividing $x^2 + 3x + 5$ by $x-1$. Compare this with $f(1)$, where $f(x) = x^2 + 3x + 5$.

16   Find the remainder on dividing $x^2 - 8x - 3$ by $x+2$. Compare this with $f(-2)$, where $f(x) = x^2 - 8x - 3$.

17   Find the remainder on dividing $x^3 - 3x^2 + x + 8$ by $x-2$. Compare this with $f(2)$, where $f(x) = x^3 - 3x^2 + x + 8$.

18   Find the remainder on dividing $2x^3 + 3x^2 - 5x + 12$ by $x+3$. Compare this with $f(-3)$, where $f(x) = 2x^3 + 3x^2 - 5x + 12$.

The division of $ax^3 + bx^2 + cx + d$ by $x - h$

Let $f(x) = ax^3 + bx^2 + cx + d$. Set out below are two computations.

A. Finding the value of $ax^3 + bx^2 + cx + d$ when x is replaced by h by the method of Section 2.

$$\begin{array}{c|cccc}
h & a & b & c & d \\
 & & ah & ah^2+bh & ah^3+bh^2+ch \\
\hline
 & a & ah+b & ah^2+bh+c & ah^3+bh^2+ch+d = f(h)
\end{array}$$

B. The long division of the same polynomial by $x-h$.

$$
\begin{array}{r}
ax^2 \qquad +(ah+b)x \qquad +(ah^2+bh+c) \\
\hline
x-h)ax^3 +bx^2 \qquad\qquad +cx \qquad\qquad +d \\
ax^3 - ahx^2 \\
\hline
(ah+b)x^2 \qquad\qquad +cx \\
(ah+b)x^2 \qquad -(ah^2+bh)x \\
\hline
(ah^2+bh+c)x \qquad\qquad +d \\
(ah^2+bh+c)x \qquad -(ah^3+bh^2+ch) \\
\hline
ah^3+bh^2+ch+d = \text{remainder}
\end{array}
$$

Comparing these two calculations, it appears that when $f(x) = ax^3 + bx^2 + cx + d$ is divided by $x-h$,

(1) the remainder is $f(h) = ah^3 + bh^2 + ch + d$
(2) the coefficients in the quotient $ax^2 + (ah+b)x + (ah^2+bh+c)$ are exactly the same numbers as occur in the lower line of the first calculation (see A).

In fact the first calculation is a very compact and schematic way of performing the division by $x-h$ and is known as *synthetic division*.

Example 1. Find the quotient and remainder on dividing $3x^3 - 5x + 10$ by $x - 2$.

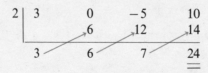

The quotient is $3x^2 + 6x + 7$ and the remainder is 24.

Example 2. Divide $x^3 + 3x^2 - 4x + 1$ by $x + 3$ and give the result in the form $f(x) = (x+3)Q(x) + R$, where $Q(x)$ is the quotient and R is the remainder.

The divisor $x+3 = x-(-3)$

$$
\begin{array}{r|rrrr}
-3 & 1 & 3 & -4 & 1 \\
   &   & -3 & 0 & 12 \\
\hline
   & 1 & 0 & -4 & 13 \\
\end{array}
$$

The quotient is $x^2 - 4$ and the remainder is 13 and so

$$x^3 + 3x^2 - 4x + 1 = (x+3)(x^2 - 4) + 13.$$

Exercise 4

As in the worked examples, use synthetic division to find the quotient and remainder when:

1    $2x^2 + 3x + 4$ is divided by $x - 1$

2    $x^2 - 4x - 8$ is divided by $x - 3$

3    $3x^2 - 5x + 7$ is divided by $x + 1$

4    $x^3 + 6x^2 + 3x + 1$ is divided by $x - 2$

5    $x^3 + 2x^2 - 3x + 4$ is divided by $x - 5$

6    $x^3 + 4x^2 - 3x - 11$ is divided by $x + 4$

7    $3x^3 + 4x^2 - 7x + 1$ is divided by $x - 3$

8    $3x^3 - 7x^2 + 5x + 4$ is divided by $x + 3$

9    $x^3 - 11x + 10$ is divided by $x + 5$

10    $x^4 - x^2 + 7$ is divided by $x + 1$

11   If $P(x) = x^3 + 2x^2 - x - 2$, show that $x + 2$ is a factor of $P(x)$. Hence factorise $P(x)$ fully.

12   Which, if any, of the linear factors $x - 1$, $x + 2$, $x + 3$ are factors of $x^3 + 2x^2 - 5x - 6$?

4   The remainder theorem

Is it true that whenever a polynomial $f(x)$ is divided by $x - h$ the remainder is $f(h)$? Questions *15–18* in Exercise 3 showed that this is true in some cases. Moreover, synthetic division does seem to show that finding the remainder by division is just the same process as computing $f(h)$. We need to be sure that the result is true in *all* cases.

When a polynomial $f(x)$ is divided by $x - h$, the quotient is another polynomial which can be denoted by $Q(x)$. The remainder R will be a constant, i.e. it will not contain x. If R contained x, the division could be carried a stage further. From Section 3, the *fundamental equation* connecting $f(x)$ with $x - h$, $Q(x)$ and R is

$$f(x) = (x - h)Q(x) + R$$

which is true for all x.

We now state a general theorem and prove it.

Theorem

If a polynomial $f(x)$ is divided by $x - h$, the final remainder is $f(h)$.

Proof. When $f(x)$ is divided by $x - h$, let $Q(x)$ be the quotient and R the remainder.

Since the degree of R is one less than that of $x - h$, R is a constant.

Now $f(x) = (x - h)Q(x) + R$ for all x (fundamental equation).

Replacing x by h,

$$\begin{aligned} f(h) &= (h - h)Q(h) + R \\ &= 0 . Q(h) + R \\ &= 0 + R \end{aligned}$$

i.e. $f(h) = R$

This result is known as the *Remainder Theorem*.

Example. Find the remainder when $x^3 - 3x + 5$ is divided by $x + 2$. Note that $x + 2 = x - (-2)$.

Method 1	*or*	Method 2

$$\begin{aligned} f(x) &= x^3 - 3x + 5 \\ \Rightarrow f(-2) &= (-2)^3 - 3(-2) + 5 \\ &= -8 + 6 + 5 \\ &= 3 \end{aligned}$$

$$\begin{array}{r|rrrr} -2 & 1 & 0 & -3 & 5 \\ & & -2 & 4 & -2 \\ \hline & 1 & -2 & 1 & 3 \end{array}$$

The remainder is 3.

Note. If the remainder only is required, the substitution method is convenient provided that the replacements for x are simple integers such as $1, -1, 2, 3$. Method 2 is generally preferable.

Exercise 5

Using the remainder theorem, find the remainder on dividing:

1	$2x^2 - 13x + 11$ by $x - 3$	*2*	$3x^2 - 5x - 3$ by $x - 2$
3	$x^3 + 4x^2 + x + 3$ by $x - 1$	*4*	$x^3 - 3x^2 + 5x - 9$ by $x - 2$
5	$2x^3 - 4x^2 + 3x - 6$ by $x - 2$	*6*	$x^3 + 4x^2 + 6x + 5$ by $x + 2$
7	$x^4 + x^2 - 16$ by $x + 1$	*8*	$x^3 - 3x^2 + 7$ by $x - 7$
9	$x^3 - x + 27$ by $x + 9$	*10*	$x^6 - x^3 - 1$ by $x - 2$

Division by $ax - b$

Since $ax - b = a\left(x - \dfrac{b}{a}\right)$, on dividing $f(x)$ by $x - \dfrac{b}{a}$ the remainder will be $f\left(\dfrac{b}{a}\right)$ and the quotient $Q(x)$.

Hence $f(x) = \left(x - \dfrac{b}{a}\right) \cdot Q(x) + f\left(\dfrac{b}{a}\right)$

$\Leftrightarrow f(x) = (ax - b) \cdot \left(\dfrac{Q(x)}{a}\right) + f\left(\dfrac{b}{a}\right)$

This shows that when $f(x)$ is divided by $ax - b$, the remainder is $f\left(\dfrac{b}{a}\right)$.

Example. Find the quotient and remainder when $f(x) = 2x^3 + x^2 + 5x - 1$ is divided by $2x - 1$.

$$
\begin{array}{c|cccc}
\tfrac{1}{2} & 2 & 1 & 5 & -1 \\
 &   & 1 & 1 & 3 \\
\hline
 & 2 & 2 & 6 & \underline{\underline{2}} = f(\tfrac{1}{2})
\end{array}
$$

$$f(x) = (x - \tfrac{1}{2})(2x^2 + 2x + 6) + 2 = (2x - 1)(x^2 + x + 3) + 2.$$

The quotient is $x^2 + x + 3$ and the remainder is 2.

Note. Remainder $= f(\frac{1}{2}) = 2(\frac{1}{2})^3 + (\frac{1}{2})^2 + 5(\frac{1}{2}) - 1 = \frac{1}{4} + \frac{1}{4} + \frac{5}{2} - 1 = 2$, which checks.

Exercise 6

Find the remainder on dividing:

1   $4x^2 + 6x - 2$ by $2x - 1$ 2 $9x^2 - 6x - 10$ by $3x + 1$

3   $6x^2 - 5x + 2$ by $3x - 1$ 4 $4x^3 - 2x^2 + 6x - 1$ by $2x - 1$

5   $2x^3 + x^2 + 4x + 4$ by $2x - 3$ 6 $2x^3 + x^2 + x + 10$ by $2x + 3$

Find the quotient and remainder on dividing:

7   $3x^3 + 5x^2 - 11x + 8$ by $3x - 1$ 8 $2x^3 + 7x^2 - 5x + 4$ by $2x + 1$

9   $2x^3 - x^2 - 1$ by $2x + 3$ 10 $5x^3 + 21x^2 + 9x - 1$ by $5x + 1$

11   $6x^3 + x^2 + 1$ by $2x - 3$ 12 $6x^2 - 11x + 8$ by $3x - 4$

Find a so that:

13   $4x^4 - 12x^3 + 13x^2 - 8x + a$ is divisible by $2x - 1$. State the quotient.

14   $6x^3 - x^2 - 9x + a$ is divisible by $2x + 3$. State the quotient.

5   *The factor theorem*

Theorem

If $f(x)$ is a polynomial, $f(h) = 0 \Leftrightarrow (x - h)$ is a factor of $f(x)$.

    Proof. From the remainder theorem $f(x) = (x - h) Q(x) + f(h)$.

         If $f(h) = 0$, then $f(x) = (x - h) Q(x)$.
         i.e. $x - h$ is a factor of $f(x)$.

    Conversely if $x - h$ is a factor of $f(x)$ then $f(x) = (x - h) Q(x)$ for some polynomial $Q(x)$.

    Hence $f(h) = (h - h) Q(h) = 0 . Q(h) = 0$.

    To sum up: $f(h) = 0 \Leftrightarrow (x - h)$ is a factor of $f(x)$.

Example. Find the factors of $2x^3 + x^2 - 13x + 6$.

Notice that if $x - h$ is a factor, with h an integer, then h divides 6. We start by trying $\pm 1, \pm 2, \pm 3, \pm 6$.

Clearly $f(1) \neq 0$ and $f(-1) \neq 0$; try $f(2)$.

$$
\begin{array}{r|rrrr}
2 & 2 & 1 & -13 & 6 \\
  &   & 4 & 10 & -6 \\
\hline
  & 2 & 5 & -3 & 0 = f(2) \\
\end{array}
$$

Since $f(2) = 0$, $x - 2$ is a factor and the other factor is $2x^2 + 5x - 3$.

$$2x^3 + x^2 - 13x + 6 = (x-2)(2x^2 + 5x - 3)$$

$$= (x-2)(2x-1)(x+3)$$

Note. Factors in which the coefficients are rational numbers are sometimes called *rational factors*, e.g. $x - 2$ and $2x - 1$, but $\sqrt{2x} - 1$ is not a rational factor. We shall be concerned mainly with rational factors.

Exercise 7

Show by means of the factor theorem that:

1   $x - 1$ and $x - 6$ are factors of $x^2 - 7x + 6$

2   $2x - 5$ is a factor of $2x^2 - 3x - 5$

3   $x - 4$ is a factor of $2x^4 - 9x^3 + 5x^2 - 3x - 4$

4   $2x - 1$ is a factor of $2x^3 + x^2 + 5x - 3$

5   $x - 1$ is a factor of $x^3 - (2a+1)x^2 + (a^2 + 2a)x - a^2$

6   Which of the following is a factor of $4x^3 - 16x^2 - x + 4$?

  a $x + 1$ *b* $x + 2$ *c* $2x - 1$

7   Which of the following are factors of $x^3 - 4x^2 - x + 4$?

  I $x - 1$ II $x - 2$ III $x + 2$ IV $x - 4$

  a   I and II only   *b*   I and III only   *c*   I and IV only   *d*   I only

Factorise fully the polynomials in questions *8–13*.

8   $x^3 - 7x + 6$ 9 $x^3 - 8x^2 + 19x - 12$

10   $x^3 - 39x + 70$ 11 $2x^3 + 7x^2 + 2x - 3$

12 $3t^3 - 4t^2 - 3t + 4$ **13** $2t^3 - 5t^2 + 4t - 21$

14   Find k for which $x^3 - 3x^2 + kx + 6$ has factor $x + 3$.

15   Find a for which $x^4 + 4x^3 + ax^2 + 4x + 1$ has factor $x + 1$.

16   Find p for which $2x^4 + 9x^3 + 5x^2 + 3x + p$ is divisible by $x + 4$. Show that when p has this value, the expression is divisible by $2x - 1$.

17   If $x + 2$ is a factor of $x^3 + kx^2 - x - 2$ find k, and the other factors when k has this value.

18   Given that $x^2 + 2x - 3$ is a factor of $f(x) = x^4 + 2x^3 - 7x^2 + ax + b$ find a and b, and hence factorise $f(x)$ completely.

19   Show that $2x + 1$ is a factor of $8x^3 + 1$, and that $8x^3 + 1$ has no other factor of the form $ax + b$, where $a, b \in R$.

20a   Divide $x^3 - h^3$ by $x - h$ and hence factorise $x^3 - h^3$.
   b   If $f(x) = x^3 + px^2 + qx + r$, find $f(x) - f(h)$.
Hence show that $x - h$ is a factor of $f(x) - f(h)$ and also that $f(x) - f(h) = (x - h)[x^2 + (p + h)x + h^2 + ph + q]$.

---

6   *Rational roots of a polynomial equation*

From Section 5:
   If $f(x)$ is a polynomial, $(x - h)$ is a factor of $f(x)$ $\Leftrightarrow$ $f(h) = 0$.
   But $f(h) = 0$ $\Leftrightarrow$ h is a root of the equation $f(x) = 0$.

It follows that if $f(x)$ is a polynomial,
$(x - h)$ is a factor of $f(x)$ $\Leftrightarrow$ h is a root of the equation $f(x) = 0$.

Example. Show that 2 is a root of the equation $x^3 - 2x^2 - x + 2 = 0$, and find the other roots. Find also the gradients of the tangents at the points where the curve $y = x^3 - 2x^2 - x + 2$ cuts the x-axis.
   Let $f(x) = x^3 - 2x^2 - x + 2$.   Dividing $f(x)$ by $x - 2$,

$$
\begin{array}{r|rrrr}
2 & 1 & -2 & -1 & 2 \\
  &   & 2 & 0 & -2 \\
\hline
  & 1 & 0 & -1 & \underline{0} = f(2). \quad f(2) = 0 \Leftrightarrow 2 \text{ is a root of } f(x) = 0.
\end{array}
$$

The equation is $(x - 2)(x^2 - 1) = 0$ $\Leftrightarrow$ $(x - 2)(x - 1)(x + 1) = 0$, with solution set $\{2, 1, -1\}$.

The curve $y = (x-2)(x-1)(x+1)$ cuts the x-axis wiᴋ giving the points of intersection $(2, 0)$, $(1, 0)$ and $(-1, 0)$.

$\dfrac{dy}{dx} = 3x^2 - 4x - 1$, so the gradient of the tangent at $(2, 0)$ is 3, at $(1, 0)$ is -2 and at $(-1, 0)$ is 6.

Exercise 8

1   Show that 1 is a root of $x^3 - 9x^2 + 20x - 12 = 0$, and find the other roots.

2   Show that -4 is a root of $6x^3 + 25x^2 + 2x - 8 = 0$, and find the other roots.

3   Show that $-\frac{1}{2}$ is a root of $4x^3 - 24x^2 + 27x + 20 = 0$, and find the other roots.

4   If 3 is a root of $x^3 - 37x + k = 0$, find k and the other roots.

Find the integral roots of the equations in questions 5–10:

5   $x^3 + 2x^2 - 5x - 6 = 0$ 6 $x^3 + x^2 - 17x + 15 = 0$

7   $x^3 - 3x + 2 = 0$ 8 $x^3 + 4x^2 + x - 6 = 0$

9   $x^4 - 1 = 0$ 10 $x^4 - 15x^2 - 10x + 24 = 0$

11   Find the rational roots of $10x^3 - 19x^2 + 9 = 0$.

12   Show that the equation $x^3 + x^2 - x + 2 = 0$ has only one real root.

13   Given that $x - 2$ is a factor of $f(x) = 2x^3 + kx^2 + 7x + 6$, find k. Hence solve the equation $f(x) = 0$ with this value of k.

14a   Find the solution set of the equation $x^3 - 6x^2 + 11x - 6 = 0$, $x \in R$.
  b   State the coordinates of the points where the curve
      $y = x^3 - 6x^2 + 11x - 6$ cuts the x-axis, and find the gradient of the curve at each of these points.

15a   Find the points where the curve $y = x^3 + x^2 - x - 1$ cuts the x-axis.
  b   Find the turning points of the curve and determine their nature.
  c   Sketch the curve, and write down the solution set of $x^3 + x^2 - x - 1 < 0$.

16   Repeat question 15 for the curve $y = x^3 - 3x - 2$.

Find the solution sets of the following equations for $0 \leqslant x \leqslant 360$:

17   $2\sin^3 x° + 3\sin^2 x° - 8\sin x° + 3 = 0$.   (Put $s = \sin x°$)

18   $6\tan^3 x° - 7\tan^2 x° - 7\tan x° + 6 = 0$.   (Put $t = \tan x°$)

7 *Approximate real roots of a polynomial equation*

When real roots of $f(x) = 0$ are not rational, rational approximations may be found by drawing the graph of $y = f(x)$ and reading off the x-coordinates of the points where the graph cuts the x-axis.

To obtain a better approximation to a root α, for example, a new graph of $y = f(x)$ is drawn in the neighbourhood of $x = \alpha$ choosing larger scale units. The process can be repeated until the root has been obtained to as many significant figures as required.

Example. Show that there is a real root of the equation $x^3 + x - 3 = 0$ between 1 and 1·5, and find an approximation for the root correct to one decimal place.

Let $f(x) = x^3 + x - 3$.
$f(1) = 1 + 1 - 3 = -1$ (negative) $\Rightarrow$ graph of $y = f(x)$ is *below* the x-axis near $x = 1$.

$$
\begin{array}{r|rrrr}
1\cdot5 & 1 & 0 & 1 & -3 \\
& & 1\cdot5 & 2\cdot25 & 4\cdot875 \\
\hline
& 1 & 1\cdot5 & 3\cdot25 & \underline{1\cdot875} = f(1\cdot5)
\end{array}
$$

$f(1\cdot5) = 1\cdot875$ (positive) $\Rightarrow$ graph of $y = f(x)$ is *above* the x-axis near $x = 1\cdot5$.

Hence the graph must cross the x-axis between $x = 1$ and $x = 1\cdot5$, and the given equation has a root α such that $1 < \alpha < 1\cdot5$, which is confirmed by the graph of $y = x^3 + x - 3$ shown in Figure 1.

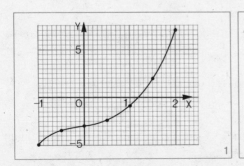

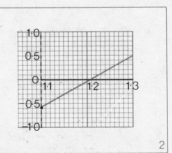

From the graph in Figure 1 we deduce that $1\cdot1 < \alpha < 1\cdot3$.

$$1\cdot1 \begin{array}{|cccc} 1 & 0 & 1 & -3 \\ & 1\cdot1 & 1\cdot21 & 2\cdot431 \\ \hline 1 & 1\cdot1 & 2\cdot21 & -0\cdot57 \doteqdot f(1\cdot1) \end{array} \qquad 1\cdot3 \begin{array}{|cccc} 1 & 0 & 1 & -3 \\ & 1\cdot3 & 1\cdot69 & 3\cdot497 \\ \hline 1 & 1\cdot3 & 2\cdot69 & 0\cdot50 \doteqdot f(1\cdot2) \end{array}$$

Using a scale of 2 cm to 0·1 unit for x and 2 cm to 1 unit for y, a first approximation to the curve for $1\cdot1 < x < 1\cdot3$ can be taken to be the coloured line joining the points $(1\cdot1, -0\cdot57)$ and $(1\cdot3, 0\cdot50)$, as in Figure 2, from which it seems likely that $1\cdot2 < \alpha < 1\cdot22$. This is in fact true since $f(1\cdot2) = -0\cdot072 \, (<0)$ and $f(1\cdot22) = 0\cdot036 \, (>0)$.

Hence $\alpha = 1\cdot2$ to one decimal place.

Exercise 9

1   Show that one root of $x^3 + x^2 + 2x - 1 = 0$ lies between 0 and 1.

2   Show that one root of $x^4 + x^2 - 1\cdot95 = 0$ lies between 0 and 1, and another between -1 and 0.

3 a Show that one root of $x^2 - 3x + 1 = 0$ lies between 0 and 0·5.

  b Draw a straight-line approximation to the graph of $y = x^2 - 3x + 1$ for $0 \leqslant x \leqslant 0\cdot5$, as shown in Figure 2. (Scales: For x, 1 cm to 0·1 unit; for y, 2 cm to 1 unit.) Hence show that the root lies between 0·38 and 0·42. State the root to one decimal place.

4 a Show that $x^3 + 2x^2 - 5 = 0$ has a root between 1 and 1·5.

  b As in question 3b find the root to one decimal place.

5   Show that $x^3 - 2x = 5$ has a root between 2 and 2·2, and find the root to one decimal place.

6 a Show that $x^3 - 3x^2 - 6 = 0$ has a root between 3·4 and 3·5.

  b Find the turning points of the curve $y = x^3 - 3x^2 - 6$, and determine the nature of each.

  c Hence sketch the curve.

7 a Show that the curve $y = x^3 - 3x^2 - 9x - 3$ cuts the x-axis between $x = -2$ and $x = -1$, between $x = -1$ and $x = 0$, and between $x = 4$ and $x = 5$.

  b Find the turning points of the curve, and determine the nature of each.

  c Sketch the curve.

8   Show that $x^3 - x^2 - 2x + 1 = 0$ has a root between 1·5 and 2, and find the root to one decimal place.

Summary

1   Polynomials

$a_n x^n + a_{n-1} x^{n-1} + \ldots + a_1 x + a_0$ is a polynomial in x of degree n, where $n \in W$ and $a_n \neq 0$.

The number a_k is called the *coefficient* of the term in x^k.

2   The division equation

When a polynomial $f(x)$ is divided by $x - h$ to give quotient $Q(x)$ and final remainder R,

$$f(x) = (x - h)Q(x) + R$$

3   Synthetic division

When $ax^3 + bx^2 + cx + d$ is divided by $x - h$, the quotient and remainder can be found from the scheme:

$$
\begin{array}{c|cccc}
h & a & b & c & d \\
 & & ah & ah^2 + bh & ah^3 + bh^2 + ch \\
\hline
 & a & ah + b & ah^2 + bh + c & \underline{ah^3 + bh^2 + ch + d} = f(h)
\end{array}
$$

This process is known as *synthetic division*.

4   The remainder theorem

If a polynomial $f(x)$ is divided by $x - h$, the remainder is $f(h)$. The remainder on division by $ax - b$ is $f(b/a)$.

5   The factor theorem

If $f(x)$ is a polynomial, $f(h) = 0 \iff (x - h)$ is a factor of $f(x)$.

6   The roots of a polynomial equation $f(x) = 0$

$(x - h)$ is a factor of $f(x) \iff h$ is a root of the equation $f(x) = 0$. Rational approximations to real roots may be found by drawing the graph of $y = f(x)$ and reading off the x-coordinates of the points where the graph cuts the x-axis. A better approximation may be obtained by drawing a part of the graph in the neighbourhood of each root choosing larger scale units. The process can be repeated until the required degree of accuracy is achieved.

The Exponential and Logarithmic Functions

1 Positive integral indices

You will recall from earlier work that x^3, or $x.x.x$, is the third power of x. The *index*, or *exponent*, 3 tells how many times the *base* x is used as a factor. In general, if a is a real number and p is a positive integer greater than 1,

$$a^p = a \times a \times \ldots \text{ to } p \text{ factors.}$$

If $p = 1$, we define $a^1 = a$.

Note also that for each positive integer p, $0^p = 0$.

In Book 7, Algebra, Chapter 2, it was shown that when a and b are non-zero real numbers, and p and q are positive integers,

(i) $a^p \times a^q = a^{p+q}$ (ii) $a^p \div a^q = a^{p-q}, p > q$

(iii) $(a^p)^q = a^{pq}$ (iv) $(ab)^p = a^p b^p$

To illustrate the essential idea in the proofs of these relationships, we now prove that (i) $a^p \times a^q = a^{p+q}$ and (ii) $a^p \div a^q = a^{p-q}, p > q$.

(i) $a^p \times a^q = (a \times a \times \ldots \text{ to } p \text{ factors}) \times (a \times a \times \ldots \text{ to } q \text{ factors})$

$\qquad = a \times a \times \ldots \text{ to } (p+q) \text{ factors}$

$\qquad = a^{p+q}$, by definition of a power.

(ii) $a^p \div a^q = \dfrac{a \times a \times \ldots \text{ to } p \text{ factors}}{a \times a \times \ldots \text{ to } q \text{ factors}}$

$\qquad = a \times a \times \ldots \text{ to } (p-q) \text{ factors, since } p > q$

$\qquad = a^{p-q}$

Examples

1. $a^4 \times a^2 = a^6$
2. $x^{10} \div x^3 = x^7$
3. $(p^2)^3 = p^6$
4. $(bc^2)^2 = b^2 c^4$

Exercise 1

Simplify:

1	$a^2 \times a^3$	*2*	$x^{10} \times x^6$	*3*	$p^2 \times p^2$	*4*	$y^{10} \times y^{40}$
5	$k^7 \div k^2$	*6*	$a^{10} \div a^5$	*7*	$b^4 \div b^3$	*8*	$x^6 \div x$
9	$(a^2)^3$	*10*	$(a^3)^2$	*11*	$(x^{10})^{10}$	*12*	$(z^2)^6$
13	$(xy^2)^3$	*14*	$2^3 \times 2^2$	*15*	$5^6 \div 5^3$	*16*	$(2^3)^2$
17	$3^4 \div 3^3$	*18*	$(a^2 b^2)^3$	*19*	$x^{12} \div x^{11}$	*20*	$a^6 \times a^6$
21	$3^3 \times 3^3 \div 3^5$	*22*	$2x^2 \times 3x^3$	*23*	$4a^6 \div 2a^3$	*24*	$(2a^2 b^3)^4$
25	$(-2)^2$	*26*	$(-2)^3$	*27*	$(-1)^4$	*28*	$(-1)^5$

29a Investigate the value of $(-1)^n$ when n is even and when n is odd. State your results.

  b If n is a positive integer, prove that $x^n - 1$ is always divisible by $x - 1$, and find the condition for which it is divisible by $x + 1$.

  c Obtain similar results for $x^n + 1$.

30 Simplify: *a* $(-x)^2$ *b* $(-x)^5$ *c* $(-2x)^3$ *d* $(-2x)^6$

31 Use the definition of a power to prove that $(ab)^4 = a^4 b^4$.

32 If a and b are real numbers and p is a positive integer, prove that $(ab)^p = a^p b^p$.

33 If a is a real number and p, q are positive integers, prove that $(a^p)^q = a^{pq}$.

---

2   Rational indices

In Book 7, Algebra, Chapter 2, we extended the meaning of powers, so that we were able to give meaning to a^p where p is not restricted to the set of positive integers.

The above laws for positive integral indices are now assumed to be true for all rational p, q. This enables us to derive meanings for powers such as a^0, a^{-p}, $a^{p/q}$.

(i) To find a meaning for a^0.

If p is a non-zero integer, $a^0 \times a^p = a^{0+p} = a^p$

so the only possible meaning for a^0 is 1.

(ii) To find a meaning for a^{-p}, $a \neq 0$.

   If p is a non-zero integer, $a^{-p} \times a^p = a^{-p+p} = a^0 = 1$ by (1),

so $a^{-p} = \dfrac{1}{a^p}$.

   Notice that in the equation $a^{-p} = \dfrac{1}{a^p}$, the indices are the negatives

of each other.

   Examples. 1. $3^4 = \dfrac{1}{3^{-4}}$ and $3^{-4} = \dfrac{1}{3^4}$ *2.* $\left(\dfrac{1}{3}\right)^{-2} = \dfrac{1}{(\frac{1}{3})^2} = \dfrac{1}{\frac{1}{9}} = 9$

Roots

If a and b are positive real numbers and n is a positive integer such that $a^n = b$, then a is defined to be the nth root of b and we write $a = \sqrt[n]{b}$ so

$$a^n = b \quad \Leftrightarrow \quad a = \sqrt[n]{b}$$

For example, $\sqrt[5]{32} = 2$ since $2^5 = 32$.

   If b is a negative real number, then an nth root of b exists provided that n is an odd positive integer. *For example,* $\sqrt[3]{(-27)} = -3$ since $(-3)^3 = -27$.

(iii) To find a meaning for $a^{3/4}$.

$$a^{3/4} \times a^{3/4} \times a^{3/4} \times a^{3/4} = a^{3/4+3/4+3/4+3/4} = a^{12/4} = a^3$$

$$\text{i.e.} \ (a^{3/4})^4 = a^3, \quad \text{so} \quad a^{3/4} = \sqrt[4]{a^3}$$

   In the same way, it can be shown that when m is an integer and n is a positive integer,

$$a^{m/n} = \sqrt[n]{a^m}.$$

For example, $8^{2/3} = \sqrt[3]{8^2} = \sqrt[3]{64} = 4$ *or* $8^{2/3} = (\sqrt[3]{8})^2 = 2^2 = 4$

Reminders

If a, b are real numbers and p, q are rational numbers,

(1) $a^p \times a^q = a^{p+q}$

(2) $a^p \div a^q = a^{p-q}$, $a \neq 0$

(3) $(a^p)^q = a^{pq}$

(4) $(ab)^p = a^p b^p$

(5) $a^0 = 1$ (6) $a^{-p} = \dfrac{1}{a^p}, a \neq 0$

(7) $a^{m/n} = \sqrt[n]{a^m}$, if m is an integer and n is a positive integer.

Exercise 2

Express with positive indices:

1 a^{-2} 2 $a^{-1/2}$ 3 $a^{-2/3}$ 4 $\left(\dfrac{1}{a}\right)^{-3}$

5 $2x^{-1}$ 6 $3x^{-1/3}$ 7 $\dfrac{3}{a^{-2}}$ 8 $\dfrac{1}{2a^{-1/2}}$

Simplify the following:

9 3^{-2} 10 $2^7 \times 2^{-5}$ 11 $2^5 \times 2^{-7}$ 12 $a^4 \times a^{-4}$

13 $(x^0)^8$ 14 $(\tfrac{1}{2})^{-2}$ 15 $\dfrac{2}{2^{-2}}$ 16 $\dfrac{3^{-2}}{3^{-4}}$

17 $4^{1/2}$ 18 $4^{-1/2}$ 19 $8^{2/3}$ 20 $27^{-2/3}$

21 $\dfrac{1}{8^{2/3}}$ 22 $\dfrac{1}{16^{-3/4}}$ 23 $\dfrac{4}{36^{-1/2}}$ 24 $\left(\dfrac{1}{9}\right)^{-1/2}$

25 $a^{1/2} \times a^{3/2}$ 26 $a^{4/3} \div a^{1/3}$ 27 $(a^{1/2})^4$

28 $a \times a^{-1/2}$ 29 $x^{1/2} \div x^{1/2}$ 30 $(a^0)^{10}$

31 $(a^{10})^0$ 32 $2x^{1/2} \times 2x^{-1/2}$ 33 $3a^{1/3} \times 4a^{1/3}$

34 $2c^{3/2} \times c^{-1/2}$ 35 $4x^{3/2} \div 2x^{1/2}$ 36 $8z^{3/4} \div 2z^{-1/4}$

Evaluate the expressions in questions 37–42 when $a = 16$ and $x = 27$:

37 $2a^{1/2}$ 38 $3a^{3/4}$ 39 $4x^{2/3}$

40 $2x^{-2/3}$ 41 $x^{1/3} \times a^{-1/4}$ 42 $3a^{-1/2} \times 4x^{-1/3}$

Simplify the following, expressing the results with positive indices:

43 $p^2 \times p^{-3}$ 44 $x^3 \div x^5$ 45 $(a^{-2})^2$

46 $2t^{1/2} \times 2t^{-3/2}$ 47 $(a^{2/3})^{-3/2}$ 48 $3x^{-3} \times 2x^2$

49 $a^{1/2}(a^{1/2} + a^{-1/2})$ 50 $(x^{1/2} - y^{-1/2})(x^{1/2} + y^{-1/2})$ 51 $(e^x - e^{-x})^2$

52 $A = 4p^2 r^3$. Calculate A when $p = 10, r = 1$.

53 $Q = 2a^{1/2}b^{-1/3}$. Calculate Q when $a = 100, b = 64$.

54a Simplify $a^{2/3} \times a^{2/3} \times a^{2/3}$, and hence show that $a^{2/3} = \sqrt[3]{a^2}$.

  b Simplify $a^{p/q} \times a^{p/q} \times \dots$ to q factors, and hence show that $a^{p/q} = \sqrt[q]{a^p}$.

---

3   The exponential function to base a, $f: x \to a^x$

Suppose that we are given a real number $a > 0$, $a \neq 1$. With each rational number x we are able to associate a real number a^x, and so we can define a function $f : x \to a^x$ which maps each rational number x on to a power of a.

Since x in a^x is an index or *exponent*, it is usual to refer to the function $f : x \to a^x$ as the *exponential function*, base a and domain Q. Can you suggest a reason why $a = 1$ is excluded as base?

As an illustration of the exponential function domain Q, take $a = 2$ so that $f : x \to 2^x$. Several values of f are given in the table below, and the graph of f corresponding to the ordered pairs in the table is shown by the dots in Figure 1.

x	$f(x) = 2^x$	
-2	2^{-2}	$= \frac{1}{4}$
-1	2^{-1}	$= \frac{1}{2}$
$-\frac{1}{2}$	$2^{-1/2}$	$\doteq 0\cdot71$
0	2^0	$= 1$
$\frac{1}{2}$	$2^{1/2}$	$\doteq 1\cdot41$
1	2^1	$= 2$
2	2^2	$= 4$
3	2^3	$= 8$

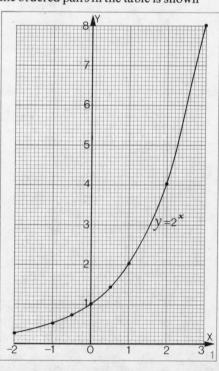

So far we have not considered 2^x for x irrational. However, since the points in Figure 1 appear to lie on a smooth curve, it will be *assumed* that there is a real number 2^x associated with each irrational number x, the value of which can be found approximately from the full graph of $f:x \rightarrow 2^x$, $x \in R$; e.g. $2^{\sqrt{3}} \doteq 2^{1 \cdot 73} \doteq 3 \cdot 3$. In Exercise 3 the exponential function is investigated for various bases.

Exercise 3

1 Draw the graph of the function $f:x \rightarrow 2^x$ for $x = -2, -1, 0, 1, 2, 3$; take a scale of 2 cm to 1 unit for x and 2 cm to 5 units for $f(x)$. Hence draw the graph of $f:x \rightarrow 2^x$ for $-2 \leqslant x \leqslant 3$, $x \in R$. From your graph, estimate to 2 significant figures

   a $2^{-1/2}$ *b* $2^{5/2}$ *c* $2^{\sqrt{2}}$ *d* $2^{\sqrt{3}}$

2 On the same sheet as for question *1*, and using the same scale, draw the graph of $g:x \rightarrow 3^x$.

   a for $x = -2, -1, 0, 1, 2, 3$ *b* for $-2 \leqslant x \leqslant 3$, $x \in R$

   What is the common point of the graphs of f and g? Why?

   Will the graph of every function $f:x \rightarrow a^x$, $a > 0$, $x \in R$, pass through this point?

3 Draw the graph of $h:x \rightarrow 10^x$ for $-2 \leqslant x \leqslant 2$, $x \in R$, choosing a suitable scale.

   a Estimate *(1)* $10^{1/2}$ *(2)* $10^{1 \cdot 2}$, to two significant figures.

   b Estimate x if *(1)* $10^x = 40$ *(2)* $10^x = 25$, to two significant figures.

4 Draw the graph of $f:x \rightarrow 1^x$, $x \in R$. Comment.

5 Sketch the graphs of $f:x \rightarrow 2^x$ and $g:x \rightarrow (\frac{1}{2})^x$ on the same diagram and with the same scales.

   a What is the axis of symmetry of the completed diagram?

   b What is the common point of the graphs of f and g?

   We shall now *assume* that the exponential function $f:x \rightarrow a^x$, $a > 0$ and $a \neq 1$, is defined for all real numbers x. The range of the function is the set of positive real numbers, which is usually denoted by R^+.

   Figure 2 shows the graph of $f:x \rightarrow a^x$ in two cases, $a > 1$ and $0 < a < 1$.

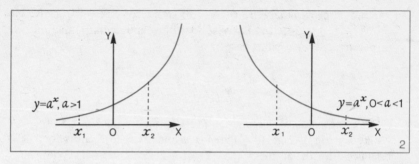

Notice that for $a > 1$, if $x_2 > x_1$, then $a^{x_2} > a^{x_1}$ so $f : x \rightarrow a^x$ is an *increasing function*, but for $0 < a < 1$, $a^{x_2} < a^{x_1}$ so f is a *decreasing function*. In both cases, $a^x > 0$ for all real numbers x, and f is one-to-one since $f(x) = f(z) \Rightarrow a^x = a^z \Rightarrow x = z$.

4 *The logarithmic function,* $f : x \rightarrow \log_a x$

Table 2

x	$f(x) = 2^x$
-2	$\frac{1}{4}$
-1	$\frac{1}{2}$
0	1
1	2
2	4
3	8

Table 1 shows some values of $f : x \rightarrow 2^x$. There is a one-to-one correspondence between the elements of the sets $\{-2, -1, 0, 1, 2, 3\}$ and $\{\frac{1}{4}, \frac{1}{2}, 1, 2, 4, 8\}$, in which x is mapped to 2^x. So there is an inverse function f^{-1}, with domain $\{\frac{1}{4}, \frac{1}{2}, 1, 2, 4, 8\}$ and range $\{-2, -1, 0, 1, 2, 3\}$ as shown in Table 2.

Table 2

x	$f^{-1}(x) = \log_2 x$
$\frac{1}{4}$	-2
$\frac{1}{2}$	-1
1	0
2	1
4	2
8	3

This inverse function is called the *logarithmic function* to base 2, and we write:

$$\log_2 \tfrac{1}{4} = -2, \qquad \log_2 \tfrac{1}{2} = -1, \qquad \log_2 1 = 0,$$
$$\log_2 2 = 1, \qquad \log_2 4 = 2, \qquad \log_2 8 = 3.$$

$\text{Log}_2 8 = 3$ is read 'The logarithm of 8 to base 2 is 3'. Notice from the corresponding entries in the two tables, that:

$$\tfrac{1}{4} = 2^{-2} \quad \Leftrightarrow \quad \log_2 \tfrac{1}{4} = -2$$
$$\tfrac{1}{2} = 2^{-1} \quad \Leftrightarrow \quad \log_2 \tfrac{1}{2} = -1$$
$$1 = 2^0 \quad \Leftrightarrow \quad \log_2 1 = 0$$
$$2 = 2^1 \quad \Leftrightarrow \quad \log_2 2 = 1$$
$$4 = 2^2 \quad \Leftrightarrow \quad \log_2 4 = 2$$
$$8 = 2^3 \quad \Leftrightarrow \quad \log_2 8 = 3$$

In general, $\qquad x = 2^n \quad \Leftrightarrow \quad \log_2 x = n$

The *logarithms* are the *indices* in the mapping $f : x \to 2^x$.

It was noted in Book 8 that if f and f^{-1} are inverse functions, the graph of one can be obtained from the graph of the other by reflecting in the line $y = x$.

In a similar way, the exponential function $f : x \to a^x$ with base a, where $a > 0$ and $a \neq 1$, has an inverse function $f^{-1} : x \to \log_a x$, called the logarithmic function to base a, and Figure 3 shows the graphs of $f : x \to a^x$ and $f^{-1} : x \to \log_a x$ for $a > 1$. The domain of f^{-1} is the set of positive real numbers, and the range is the set of real numbers.

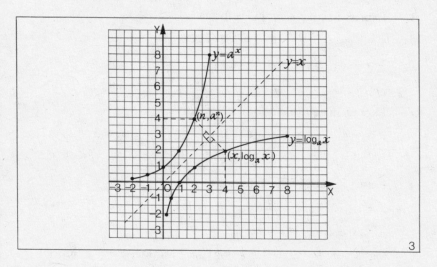

3

Now under reflection in the line $y = x$, the point $(x, y) \to (y, x)$. Hence if (n, a^n) and $(x, \log_a x)$ are corresponding points,

$$x = a^n \quad \Leftrightarrow \quad \log_a x = n$$

Note (1) $1 = a^0 \quad \Leftrightarrow \quad \log_a 1 = 0$
 (2) $a = a^1 \quad \Leftrightarrow \quad \log_a a = 1$

Example 1.

(i) $100 = 10^2 \Leftrightarrow \log_{10} 100 = 2$ (ii) $\frac{1}{25} = 5^{-2} \Leftrightarrow \log_5 \frac{1}{25}$

Example 2. Find $\log_3 9$.

Let $\log_3 9 = n$.
Then $9 = 3^n \Leftrightarrow 3^2 = 3^n$, from which $n = 2$.

Exercise 4

1   Draw up tables like the ones on page 42 to show values of f defined by $f(x) = 10^x$ for $-2 \leqslant x \leqslant 4$, and of $f^{-1}(x) = \log_{10} x$. Hence express the following in equivalent logarithmic form:

$$\tfrac{1}{100} = 10^{-2}, \quad 1 = 10^0, \quad 100 = 10^2, \quad 10\,000 = 10^4.$$

2   Express in equivalent logarithmic form:

a $8 = 2^3$ b $4 = 2^2$ c $1 = 2^0$ d $\frac{1}{16} = 2^{-4}$ e $p = 2^q$

3   Express in equivalent index form:

a $4 = \log_2 16$ b $7 = \log_2 128$ c $\log_5 125 = 3$

d $-6 = \log_2 \frac{1}{64}$ e $1 = \log_7 7$ f $\log_2 a = 5$

g $\log_3 4 = x$ h $\frac{1}{3} = \log_2 \sqrt[3]{2}$ i $\log_a n = b$

4   Express in equivalent logarithmic form:

a $9 = 3^2$ b $64 = 4^3$ c $\frac{1}{9} = 3^{-2}$

d $1 = 9^0$ e $\sqrt[3]{5} = 5^{1/3}$ f $27^{-1/3} = \frac{1}{3}$

5   Find x in each of the following:

a $x = \log_2 2$ b $-2 = \log_4 x$ c $\log_2 x = 5$ d $x = \log_3 \frac{1}{27}$

6   Find the values of:

a $\log_9 81$ b $\log_6 6$ c $\log_2 1$ d $\log_3 \frac{1}{9}$ e $\log_{81} 9$

7   *Estimate*, to two significant figures where possible, the values of:

a $\log_3 10$ b $\log_2 12$ c $\log_{10} 80$ d $\log_5 100$ e $\log_9 4$

5   The laws of logarithms

When the index laws are expressed in logarithmic form, they are often referred to as the *laws of logarithms*. We have made use of them in calculations from Book 5 onwards.

If a, x and y are positive real numbers, and $a \neq 1$, then

1. $\log_a xy = \log_a x + \log_a y$ 2. $\log_a \dfrac{x}{y} = \log_a x - \log_a y$

Proofs. Let $\log_a x = m$ and $\log_a y = n$, so that $x = a^m$ and $y = a^n$.

1. $xy = a^m \times a^n = a^{m+n}$ 2. $\dfrac{x}{y} = a^m \div a^n = a^{m-n}$

$\Leftrightarrow \log_a xy = m+n$ $\Leftrightarrow \log_a \dfrac{x}{y} = m-n$

$\Leftrightarrow \log_a xy = \log_a x + \log_a y$ $\Leftrightarrow \log_a \dfrac{x}{y} = \log_a x - \log_a y$

Note. The base is sometimes omitted from the notation when it is obvious from the context, or in statements which are true for logarithms to every base.

Example 1. $\log 5 + \log \frac{1}{5}$
$= \log(5 \times \frac{1}{5})$
$= \log 1$
$= 0$ (for every base)

Example 2. $\log_2 4 + \log_2 12 - \log_2 6$
$= \log_2 \left(\dfrac{4 \times 12}{6} \right)$
$= \log_2 8$
$= 3$

Exercise 5

1   Simplify:

  a $\log_{10} 5 + \log_{10} 2$ b $\log_2 16 - \log_2 4$ c $\log_5 5 + \log_5 5$

2   Simplify:

  a $\log_6 9 + \log_6 8 - \log_6 2$ b $\log_{10} 2 + \log_{10} 10 - \log_{10} \frac{1}{5}$

  c $\log 3 + \log 4 + \log \frac{1}{2} + \log \frac{1}{6}$ d $\log_2 2 + \log_2 3 - \log_2 6 - \log_2 8$

3   Which of the following are true and which are false, for logarithms to every base?

a $\log 2 + \log 3 = \log 5$

b $\log 3 + \log 4 = \log 12$

c $\log 8 = \log 4 + \log 2$

d $\log 10 + \log 10 = \log 100$

e $\log 2 \times \log 3 = \log 6$

f $\log 8 \div \log 2 = \log 4$

g $\log 3^2 + \log 3^{-2} = 0$

h $\log \dfrac{5}{3} = \dfrac{\log 5}{\log 3}$

4   Verify that:

a $\log 3 + \log \frac{1}{3} = \log 1$

b $\log 4 + \log \frac{1}{4} = 0$

c $\log \frac{1}{5} = -\log 5$

d $\log \dfrac{1}{c} = -\log c,\ c > 0$

5   Simplify:

a $\log_2 15 + \log_2 14 - \log_2 105$

b $\log_{10} 100 + \log_{10} 0{\cdot}1$

c $\log_3 9 - \log_3 \frac{1}{3}$

d $\log_a 1 + \log_a 1 + \log_a 1$

e $\log_2 \frac{1}{8} - \log_2 \frac{1}{64}$

f $\log_2 (2^8) + \log_2 (\frac{1}{8})^2$

6   Solve for x, where x is a positive real number:

a $\log x + \log 3 = \log 6$

b $\log x - \log 15 = \log 0{\cdot}2$

c $\log (x+1) + \log (x-1) = \log 3$

d $\log x^2 + \log 0{\cdot}1 = \log 10$

e $\log_5 (x+1) + \log_5 (x-3) = 1$

f $\log_2 (3x-5) - \log_2 (x+2) = 1$

7   Given $\log_a 8 + \log_a 4 - \log_a 2 = 2$, find a.

8   Verify that:

a $\log 3 + \log 3 = \log 3^2$

b $\log 4 + \log 4 + \log 4 = \log 4^3$

c $\log 5^4 = 4 \log 5$

d $\log a^n = n \log a,\ a > 0$

9   If $y = \log_e x - \log_e (1+x),\ x > 0$, show that $x = \dfrac{e^y}{1-e^y}$.

3. $\log_a x^n = n \log_a x,\ x > 0,\ n$ rational

Proof. Let $\log_a x = m$, so that $x = a^m$.
$$x^n = (a^m)^n = a^{mn}$$
$$\Leftrightarrow \log_a x^n = nm$$
$$\Leftrightarrow \log_a x^n = n \log_a x$$

Example 1. Simplify $\log x^3 + \log \dfrac{1}{x} - 2 \log x$, $x > 0$.

$\log x^3 + \log \dfrac{1}{x} - 2 \log x$ *or* $\log x^3 + \log \dfrac{1}{x} - 2 \log x$

$= 3 \log x + \log 1 - \log x - 2 \log x$ $= \log \left(x^3 \times \dfrac{1}{x} \div x^2 \right)$

$= 3 \log x + 0 - 3 \log x$ $= \log 1$

$= 0$ $= 0$

Example 2. Solve the equation $2^x = 6$, $x \in R$.

Taking logarithms to base 10 of both sides,

$\log_{10} 2^x = \log_{10} 6$

$\Leftrightarrow \quad x \log_{10} 2 = \log_{10} 6$

$\Leftrightarrow \qquad x = \dfrac{\log_{10} 6}{\log_{10} 2}$

$\qquad\qquad = \dfrac{0 \cdot 778}{0 \cdot 301}$

$\qquad\qquad = 2 \cdot 6$, to two significant figures.

nos	*logs*
0·778	$\bar{1}$·891
0·301	$\bar{1}$·479
2·58	0·412

Exercise 6

1   Simplify:

   a $2 \log_{10} 5 + 2 \log_{10} 2$ *b* $\frac{1}{2} \log_2 16 - \frac{1}{3} \log_2 8$ *c* $2 \log x - \log x^2$

2   Simplify:

   a $\log 2 + 2 \log 3 - \log 18$ *b* $2 \log_4 2 - 2 \log_4 4 - \log_4 \frac{1}{4}$

3   *a*   Show that $\log 81 = 4 \log 3$

   b   Hence simplify: *(1)* $\log 81 - \log 27$ *(2)* $\dfrac{\log 81}{\log 27}$

4   Which of the following are true and which are false, for logarithms to every base?

   a $\log 5^{-2} = -2 \log 5$ *b* $\log 4 = \frac{2}{3} \log 8$

   c $\log 27 = \frac{3}{4} \log 81$ *d* $\frac{1}{3} \log 11 = \log \frac{11}{3}$

   e $\log 5 = \frac{1}{2} \log 10$ *f* $\log 2 - \log \sqrt{2} = \log \sqrt{2}$

   g $\log \frac{1}{5} - \log 5 = -\log 25$ *h* $\dfrac{\log \sqrt{2}}{\log \sqrt{8}} = \dfrac{1}{3}$

5 Which is the greatest of $\frac{2}{3}\log 1$, $\frac{3}{4}\log 1$, $\frac{4}{3}\log 1$?

6 Simplify:

a $\log x^4 - 3\log x + \log \dfrac{1}{x}$

b $\log x^{1/2} + \log y^{1/2} - \frac{1}{2}\log xy$

c $\log_2 \sqrt{6} - \frac{1}{2}\log_2 3$

d $\frac{1}{2}\log_{10} 10 + 3\log_{10}\sqrt{10}$

7 Solve the following equations, where $x \in R$. Round off the roots to two significant figures.

a $3^x = 7$ b $8^x = 5$ c $12^x = 981$ d $2^x = 1\cdot25$

8 Find the least positive integer n for which

a $3^n > 6090$ b $\left(\dfrac{1}{3}\right)^n < \dfrac{1}{10^5}$

9 How many terms of the geometric series $100 + 50 + 25 + \ldots$ must be taken for the sum to exceed 199?

10 Use tables of logarithms to evaluate:

a $2^{2\cdot5}$ b $3^{3\cdot1}$ c $1\cdot05^{1\cdot2}$ d $0\cdot2^{1/3}$

11 Use tables of logarithms to find which member of each pair is the larger:

a 2^{17} or 3^{11} b 15^6 or 16^5

12 Prove that if $a_1, a_2, a_3, \ldots, a_n$ is a geometric sequence of positive terms, then $\log a_1, \log a_2, \log a_3, \ldots, \log a_n$ is an arithmetic sequence.

13 By taking logarithms of both sides to base 10, express the following formulae in a form suitable for logarithmic calculations:

a $A = 4\pi r^2$ b $V = \frac{4}{3}\pi r^3$ c $T = 2\pi\sqrt{\dfrac{l}{g}}$ d $pv^{1\cdot4} = c$

14 If $2\log_3 y = \log_3(x+1) + 2$, show that $y^2 = 9(x+1)$.

6   Change of base of logarithms

It is sometimes useful to be able to find $\log_b x$ when logarithms to base a are known, for example to be able to find $\log_2 x$ when tables of $\log_{10} x$ are available. We show that:

$$\log_b x = \frac{\log_a x}{\log_a b}.$$

Proof. Let $\log_b x = m$, so that $x = b^m$.

Taking logarithms to base a of both sides,

$$\log_a x = \log_a b^m = m \log_a b$$

$$\Leftrightarrow \quad m = \frac{\log_a x}{\log_a b}$$

$$\Leftrightarrow \log_b x = \frac{\log_a x}{\log_a b}$$

Example. Given that $e \doteqdot 2.72$, calculate $\log_e 8$.

$$\log_e 8 = \frac{\log_{10} 8}{\log_{10} e} = \frac{\log_{10} 8}{\log_{10} 2.72} = \frac{0.903}{0.435} \doteqdot 2.1$$

nos	logs
0·903	$\overline{1}$·956
0·435	$\overline{1}$·638
2·08	0·318

Note. Two bases for logarithms are in general use:

(i) Common logarithms, using base 10.
(ii) Natural, or Napierian, logarithms using base $e \doteqdot 2.72$.

Exercise 7

Use common logarithms to calculate the following, rounding off answers to two significant figures. Take $e \doteqdot 2.72$.

1	$\log_2 7$	*2*	$\log_e 10$	*3*	$\log_3 5$	*4*	$\log_5 3$
5	$\log_e \pi$	*6*	$\log_2 e$	*7*	$\log_{20} 100$	*8*	$\log_{100} 20$

9   Replacing x by a in the equation $\log_b x = \dfrac{\log_a x}{\log_a b}$, prove that

$\log_b a \times \log_a b = 1$.

10   Check the result in question **9** for $a = 9$ and $b = 4$.

11   Given $\log_e 10 = 2{\cdot}30$, show that $\log_{10} e = 0{\cdot}435$.

12   If $\log_a 3 = \log_b 27$, show that $a^3 = b$.

13   Show that for any positive number N, $\log_e N \doteqdot 2{\cdot}30 \log_{10} N$, and use this formula to calculate $\log_e 6{\cdot}31$.

14   Calculate: a $\log_{1/2} 5$ b $\log_5 \frac{1}{2}$.
   (Note that $\bar{1}{\cdot}699 = -1 + 0{\cdot}699 = -0{\cdot}301$.)

---

7   *The derivation from experimental data of a law of the form* $y = kx^n$

In Book 6, Algebra, Chapter 2, we saw how to derive certain formulae from experimental data. The method was to test graphically for a linear relation connecting sets of corresponding readings of two variables x and y. For example, if the formula was thought to be of the form $y = ax + b$, we plotted y against x. If the points lay on a straight line, then the relation was indeed given by $y = ax + b$ since we know that every straight line has an equation of the form $y = mx + c$.

The constants a and b were found by substituting in $y = ax + b$ the coordinates of two well-separated points on the *best-fitting straight line*, and solving the resulting system of equations for a and b.

The method can be applied to laws of the form $y = kx^n$, where k and n are constants to be determined. If x and y are variables related by the equation $y = kx^n$, then

$$\log y = \log kx^n = \log k + \log x^n$$

i.e. $\log y = n \log x + \log k$

Putting $\log y = Y$, $\log x = X$, $n = a$ and $\log k = b$, this becomes

$$Y = aX + b,$$

which is the equation of a straight line.

Conversely, suppose $Y = \log y$ plotted against $X = \log x$ gives a straight line $Y = aX + b$, within the limits of experimental error, then putting $a = n$ and $b = \log k$, we have

$$\log y = n \log x + \log k$$
$$\Leftrightarrow \log y = \log kx^n$$
$$\Leftrightarrow \quad y = kx^n.$$

It is usually convenient to take logarithms to base 10, using tables or a slide rule (L scale).

Example. The readings given were obtained in an experiment. Show that the variables are related by an equation of the form $y = kx^n$, and find k and n.

x	20	25	32	40	50
y	35	55	76	110	158

From tables of logarithms we have to 2 decimal places,

$\log_{10} x$	1·30	1·40	1·50	1·60	1·70
$\log_{10} y$	1·54	1·74	1·88	2·04	2·20

Allowing for experimental error, the graph of $\log_{10} y$ against $\log_{10} x$ is approximately a straight line, as shown in Figure 4(i).

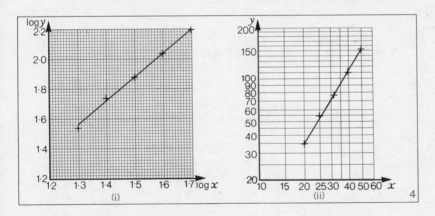

Hence its equation is of the form $Y = aX + b$,

i.e. $$\log_{10} y = a \log_{10} x + b,$$

and so x and y are related by a formula $y = kx^n$, where $\log_{10} k = b$ and $n = a$.

To estimate a and b we take two well-separated points on the straight line, for example $(1·70, 2·20)$ and $(1·30, 1·56)$. Substituting these coordinates in the equation of the line gives the equations

$$2·20 = a \times 1·70 + b . \qquad . \qquad . \qquad . \qquad (1)$$
$$1·56 = a \times 1·30 + b . \qquad . \qquad . \qquad . \qquad (2)$$

Subtract $\qquad\qquad\qquad \overline{0·64 = 0·4a}$

$\qquad\qquad \Leftrightarrow \qquad a = 1·6$

From equation (1), $\qquad b = -0·52.$

Now $n = a = 1·6$, and

$\log_{10} k = b = -0·52 = \bar{1}·48$, from which $k = 0·3.$

So the formula $y = kx^n$ becomes $y = 0·3x^{1·6}.$

Note. (i) The two points selected must lie on the line, and be well separated.

(ii) Figure 4(ii) shows the same result plotted on log-log graph paper. On this paper the readings can be plotted directly to give a straight-line graph, since the scales convert x to $X = \log x$, y to $Y = \log y$. Compare with scales on a slide rule.

Exercise 8

1 The results of an experiment to find the relation between two variables x and y give this table. Draw the graph of $\log_{10} y$ against $\log_{10} x$. Explain why the relation is of the form $y = kx^n$, and find k and n.

$\log_{10} x$	0·1	0·2	0·3	0·4	0·5
$\log_{10} y$	0·60	0·90	1·25	1·50	1·80

2 The table shows the logarithms of corresponding readings of x and y in an experiment. Show that $y = kx^n$, and find k and n.

$\log_{10} x$	1·70	2·29	2·70	2·85	3·00
$\log_{10} y$	1·32	1·67	1·92	2·01	2·10

3   Construct a table of values of $\log_{10} x$ and $\log_{10} y$ from the data given below. By drawing an appropriate graph show that x and y are related by an equation of the form $y = kx^n$. Find the relation.

x	1	4	9	16
y	4·0	2·0	1·3	1·0

4   Plot $\log_{10} E$ against $\log_{10} p$. Hence show that $E = kp^n$ and find k and n.

p	14·1	28·2	63·1	126
E	15·9	6·31	3·16	1·58

5   Repeat question 4 for this table of readings.

p	1	1·40	1·91	2·82	3·98
E	0·5	1·0	2·0	3·55	7·95

6   The table shows corresponding measurements of the diameter (d cm) and the breaking load (m kg) of a rope. Show that the diameter and breaking load are related by a law of the form $m = kd^n$, and find k and n.

d	1·0	2·0	3·0	4·0	5·0
m	19	80	177	316	500

8   Exponential growth and decay

There is evidence of growth and decay all around us in the world. Plants grow and decay, populations increase and decrease, radioactive elements disintegrate and decay, money left in a savings account in the bank increases with interest, and so on.

A particularly interesting type of change is that where

$$(\text{rate of change of size}) \propto \text{size}$$

For example £100 left in a savings bank at compound interest will grow at four times the rate of £25 left in the same bank under the same conditions. In the time £25 takes to grow to £50, £100 will grow to £200.

Figure 5 shows the growth of £100 at 10% per annum compound interest over a period of 25 years. The amount at the end of each year is shown by a dot; if interest were added continuously we would obtain the smooth *exponential* curve shown. (Compare Figure 1 on page 40.)

In Book 8 we saw that the formula for calculating the amount £A of principal £P at rate r% per annum after n years was

$$A = P\left(1 + \frac{r}{100}\right)^n = PR^n,$$

where $R = 1 + \dfrac{r}{100}$.

This is of the form $y = ka^x$.

All exponential growth and decay can be represented by a formula of this type.

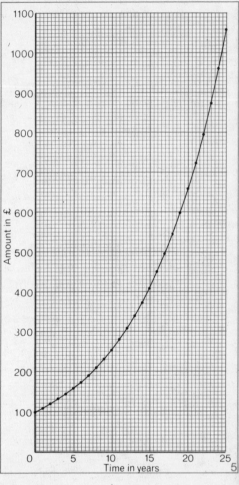

Examples of this type of growth and decay include the growth of living organisms, the change in temperature of a cooling body and the rate of a chemical reaction.

In practice the base a is often chosen to be 2·7183..., which is denoted by e, the base of *natural logarithms*. Values of $\log_e x$, and of e^x (the exponential function) are given in tables and slide rules (LL scales).

Example. The intensity I_0 of a light source is reduced to I after passing through d metres of a fog according to the formula $I = I_0 e^{-0.14d}$. In what distance will the intensity be reduced to 0·01 of its original value?

Here $I = 0.01I_0$, so we have

$$0.01I_0 = I_0 e^{-0.14d}$$

$$\Leftrightarrow \quad 0.01 = e^{-0.14d}$$

Taking logarithms to base 10,

$$\log 0.01 = -0.14d \log e$$

$$\Leftrightarrow \quad d = \frac{\log 0.01}{-0.14 \log 2.72}$$

$$= \frac{-2}{-0.14 \times 0.435}$$

$$= 32.9$$

nos	logs
2	0·301
0·14	$\bar{1}$·146
0·435	$\bar{1}$·638
	$\bar{2}$·784
32·9	1·517

So the distance is approximately 33 metres.

Exercise 9B

1   If a sum £P is invested at r per cent per annum compound interest, then the amount £A at the end of $\dot{n}$ years is given by $A = P\left(1 + \dfrac{r}{100}\right)^n$. Estimate A when $P = 500$, $r = 8$ and $n = 10$.

2   Using the compound interest formula in question *1*,

   a find P when $A = 400$, $r = 10$ and $n = 5$
   b in what time will £P be doubled when $r = 6$?

3   A current I_0 amperes falls to I amperes after t seconds according to the formula $I = I_0 e^{-kt}$. Find the constant k if a current of 10 amperes falls to 1 ampere in 0·01 second.

4   A radioactive mineral decays according to the formula $m = m_0 e^{-0.05t}$, where m_0 is the initial mass and m is the mass after t years.

   a Find m for $m_0 = 500$ and $t = 50$.
   b The time for which $m = \frac{1}{2}m_0$ is the half-life of the substance. Find this time.

5   An X-ray beam of intensity I_0 in passing through absorbing material x millimetres thick emerges with an intensity I given by $I = I_0 e^{-\mu x}$. It is found that I is 50% of I_0 when $x = 9$. Calculate the constant μ.

6   A radioactive element disintegrates according to the formula $m = m_0 e^{-kt}$, where m_0 is the initial mass and m is the mass after t seconds.

    a If at the end of one second half of the material has disintegrated, show that $e^{-k} = \frac{1}{2}$.

    b Show that if two-thirds of the material has disintegrated after T seconds, $\frac{1}{3} = (\frac{1}{2})^T$. Hence find T.

7   The atmospheric pressure P at height h kilometres is given by $P = P_0 e^{-kh}$. The pressure at sea-level, P_0, is 760 units.

    a If $P = 670$ when $h = 1$, show that $e^{-2k} = (\frac{67}{76})^2$.

    b Calculate the pressure at a height of 2 km.

8   The population of a certain town is 30 000. Assuming that the growth is exponential, the population P after t years is given by the formula $P = 30\,000 \times 1.2^{0.1t}$.

    a Estimate the population after 25 years.

    b How long will it take the population to double?

9   A hot piece of metal loses heat according to the formula $T = T_0 e^{-0.2t}$, where T is the temperature difference between the metal and the surrounding air after t minutes, and T_0 is the initial temperature difference.

    Initially the temperature of the metal was 330°C, and of the air 30°C. Find the temperature of the metal after:

    a 5 minutes   *b* 20 minutes.

10   A machine, costing £P initially, depreciates each year by $\dfrac{1}{r}$ of its estimated value at the end of the previous year. Its estimated value £V after n years is given by the formula $V = P\left(1 - \dfrac{1}{r}\right)^n$.

    a Find the estimated value after 3 years of a machine which initially cost £9600, taking $r = 4$.

    b If $V = 1000, P = 5000$ and $r = \frac{1}{3}$, find n to the nearest whole number.

Summary

Indices
With the necessary restrictions on a, b, m, n, p, q.

1   $a^p = a \times a \times \ldots$ to p factors.

2   $a^p \times a^q = a^{p+q}$; $a^p \div a^q = a^{p-q}$
    $(a^p)^q = a^{pq}$; $(ab)^n = a^n b^n$

3   $a^0 = 1$; $a^{-p} = \dfrac{1}{a^p}$; $a^{m/n} = \sqrt[n]{a^m} = (\sqrt[n]{a})^m$

Exponential and logarithmic functions
4   The exponential function $f : x \rightarrow a^x$, and the

    logarithmic function $l : x \rightarrow \log_a x$ are inverse functions.

$$x = a^n \iff \log_a x = n$$

Logarithms
5   $\log xy = \log x + \log y$; $\log \dfrac{x}{y} = \log x - \log y$; $\log x^n = n \log x$

6   $\log_b x = \dfrac{\log_a x}{\log_a b}$ (change of base formula).

Linear graph
7   The graph of $\log y$ against $\log x$ is a straight line $\iff y = kx^n$.

Exponential growth and decay
8   A formula for exponential growth and decay: $y = ka^x$.

Pattern and Structure: Groups

1 Operations

Starting from an element such as a number, a figure or a set, a *unary* operation produces a unique single element. For example, the operation "Add 5" applied to a number such as 31 produces a unique number, in this case 36. See Figure 1.

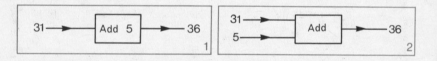

In this chapter we shall be concerned with *binary* operations. A binary operation starts with *two* elements in a given order and produces from them a unique single element. Compare Figure 2 with Figure 1.

Exercise 1

1 Consider the operations indicated, and decide whether they are unary or binary. Give examples of the kind of element or elements that they might operate on (e.g. numbers, matrices, vectors, sets).

a Doubling   *b* +   *c* ×   *d* reflecting a plane figure in a given line

e ∩   *f* taking the complement of a set   *g* − as in '*a* − *b*'

2 Which of the following symbols usually represent a binary operation? For those which do, state what kind of element they produce.

a ÷   *b* >   *c* ∪   *d* ⊂

e ∘, as used in composition of functions or transformations.

2 Cayley tables

In many cases the simplest way to show the elements produced by a *binary* operation is by a *two*-dimensional array. You have already met examples such as multiplication tables for natural numbers, addition tables for the integers and composition tables for the composition of functions and transformations. A general name sometimes used for all of these is *Cayley tables* after the British mathematician Cayley (1821–1895) who invented them in 1854.

Exercise 2

Copy and complete the following tables, or in the case of infinite arrays indicate how the table could be continued. *Keep the tables for future use.*

1

+	E	O
E		
O		E

E denotes an even number.
O denotes an odd number.
+ denotes ordinary addition.

2

∘	D	O
D		
O		

D denotes a direct geometrical transformation.
O denotes an opposite geometrical transformation.
∘ as used in composition of transformations.

3

+	0	1	2	3	...
0				3	...
1					
2					
3	3				
⋮	⋮				

$\{0, 1, 2, 3, ...\}$ is the set of whole numbers.
+ is ordinary addition.

4

$+_3$	0	1	2
0			
1			
2			

$+_3$ denotes addition in 3-hour clock arithmetic (based on a clock with the numerals 0, 1, 2).

5

∘	I	R	R²
I			
R			
R²		R	

I is the identity transformation.

R, R^2 denote anticlockwise rotations about O of $120°$ and $240°$.

∘ as used in composition of transformations.

6

×	I	A	A²
I	I		
A		A²	
A²			

$$I = \begin{pmatrix} 1 & 0 \\ 0 & 1 \end{pmatrix}, \quad A = \begin{pmatrix} -\tfrac{1}{2} & -\tfrac{1}{2}\sqrt{3} \\ \tfrac{1}{2}\sqrt{3} & -\tfrac{1}{2} \end{pmatrix},$$

$A^2 = A \times A$.

× denotes matrix multiplication.

7

×	1	2	3	4	...
1				4	...
2					
3					
4	4				
⋮					

$\{1, 2, 3, 4, \dots\}$ is the set of natural numbers.

× is ordinary multiplication.

8

×₅	1	2	3	4
1	1	2	3	4
2	2			
3	3			
4	4			

$\times_5$ is multiplication in 5-hour clock arithmetic. (Notice that the entries are the *remainders* on division by 5.)

9

+₄	0	1	2	3
0				
1				
2				
3				

$+_4$ is addition in 4-hour clock arithmetic.

10

⊕	**0**	**a**	**b**	**c**	...
0	**0**	**a**	**b**	**c**	...
a	**a**	**2a**	**a ⊕ b**	**a ⊕ c**	
b					
c					
⋮					

0, a, b, c are vectors.

⊕ denotes vector addition.

11

×	A	B	C	D
A				
B				
C				
D				

$$A = \begin{pmatrix} 1 & 0 \\ 0 & 1 \end{pmatrix}, \quad B = \begin{pmatrix} -1 & 0 \\ 0 & -1 \end{pmatrix},$$

$$C = \begin{pmatrix} 1 & 0 \\ 0 & -1 \end{pmatrix}, \quad D = \begin{pmatrix} -1 & 0 \\ 0 & 1 \end{pmatrix}.$$

× is matrix multiplication.

12

×	P	Q	R	S
P				
Q				
R				
S				

$$P = \begin{pmatrix} 1 & 0 \\ 0 & 1 \end{pmatrix}, \quad Q = \begin{pmatrix} 0 & -1 \\ 1 & 0 \end{pmatrix},$$

$$R = \begin{pmatrix} -1 & 0 \\ 0 & -1 \end{pmatrix}, \quad S = \begin{pmatrix} 0 & 1 \\ -1 & 0 \end{pmatrix}.$$

× is matrix multiplication.

13

∪	∅	A	B	C
∅	∅			
A				C
B		C		
C				C

$A = \{1, 2, 3\}, B = \{3, 4, 5\},$
$C = \{1, 2, 3, 4, 5\}.$
∪ is the operation of union.

14

∩	∅	A	B	C
∅				
A			{3}	
B				
C				

A, B, C are the sets in question **13**.
∩ is the operation of intersection.

15

	First mapping		
∘	f	g	h
f			
g		*	
h		g	

(Second mapping labels the rows)

$f : x \to x,$

$g : x \to 1 - \dfrac{1}{x},$

$h : x \to \dfrac{1}{1-x}.$

∘ denotes composition of functions.

Note that since $(g \circ h)(x) = g(h(x))$, the first mapping in $g \circ h$ is h. Hence at * the entry is $g \circ h$, i.e. f.

---

3 Closure, identity, inverses

Each of the tables in Exercise 2 shows a *system* consisting of a *set*, either finite or infinite, together with a *binary* operation. Notice the wide variety of the systems. The interest and importance of the abstract pattern of structure we discuss comes from its wide-ranging applicability. In each of the examples above, some or all of the following properties can be found:

(i) *Closure.* The system is closed, i.e. the operation always produces an element of the set.

For example, in question *4* the operation always produces an element of the original set $\{0, 1, 2\}$.

(ii) *Identity*. The system has an identity element.

For example, in question *10* the identity element is the zero vector *0*, since $a \oplus 0 = a = 0 \oplus a$ for all *a*.

In question *15* the identity element is f since $f \circ f = f, f \circ g = g \circ f = g$, and $f \circ h = h \circ f = h$.

(iii) *Inverses*. Each element in the system has an inverse in the system.

For example, in question *15* the inverse of g is h so that $g \circ h = h \circ g = f$, and the inverse of h is g.

Exercise 3

1   Which of the systems in Exercise 2 have the closure property?

2   Every system in Exercise 2 has an identity element. Name the identity element in each case.

3   Which of the systems in Exercise 2 have the inverse property, i.e. an inverse for *every* element of the system?

---

4   Associativity

If a, b, c are members of a set S with an operation $*$ then it may or may not be true that the associative property holds for this operation, i.e. that

$$a * (b * c) = (a * b) * c \quad \text{for all} \quad a, b, c \in S.$$

We are so accustomed to working with associative operations, such as addition and multiplication of numbers, union and intersection of sets, and the composition of functions and transformations, that we tend to expect all operations to be associative.

Exercise 4

1 By producing one counter-example in each case, show that none of the following are associative:

  a operation $-$ on Z　　　b operation $\div$ on set $\{x : x \in Q \text{ and } x > 0\}$

  c operation $*$ on N, where $a * b = a^b$

2 For an associative operation, e.g. $+$, $a + (b + c)$ can be written without brackets, unambiguously. What meaning, if any, is conventionally given to:

  a $p - q - r$　　　　　b $a \div b \div c$　　　　　c x^{y^z} ?

5　Groups

If a system A, $*$, consisting of a set A and a binary operation $*$, is *closed*, *associative* and has an *identity* element, and if each element has an *inverse* in A, then the system is called a *group*.

Exercise 5

1 Which of the systems in Exercise 2 are groups?

  State whether each of the following satisfies all four requirements for a group. If it does not, say why not.

2 $N, +$　　　3　$Z, +$　　　4　$Z, \times$　　　5　$Q, \times$

6 non-zero rational numbers, $\times$

7 the set of all functions: $R \to R$, $\circ$

8 the set of rotations in a plane about a given fixed point, $\circ$

9 the set of 2×2 matrices, matrix addition

10 the set of even integers, $+$

11 the set of odd integers, $+$

12 the set of 2×2 matrices, matrix multiplication

13 subsets of set E, $\cap$.

Example. Explain the solution of the equations $3+x = 5$ and $3x = 5, x \in R$.

From earlier work in the course (summed up in the table of structure in the chapter on *Number Systems and Surds* in Book 7), we now know that:

(i) the set of real numbers under addition is a group
(ii) the set of non-zero real numbers under multiplication is a group.

So we have:

$3+x = 5$	*x a variable on R*	$3x = 5$
$-3+(3+x) = -3+5$	*existence of inverse*	$\frac{1}{3}(3x) = \frac{1}{3} \times 5$
$-3+(3+x) = 2$	*closure property*	$\frac{1}{3}(3x) = \frac{5}{3}$
$(-3+3)+x = 2$	*associative property*	$(\frac{1}{3} \times 3)x = \frac{5}{3}$
$0+x = 2$	*inverse property*	$1 \times x = \frac{5}{3}$
$x = 2$	*identity property*	$x = \frac{5}{3}$

Exercise 6

Can the following equations always be solved for X? Give reasons for your answers.

1 $R \circ X = Q$, where $X, R, Q \in \{$rotations in a plane about the origin$\}$

2 $A \cap X = B$, where $X, A, B \in \{$subsets of $E\}$

3 $P+X = M$, where $X, P, M \in \{2 \times 2$ matrices$\}$

4 $PX = M$, where $X, P, M \in \{2 \times 2$ matrices$\}$

---

6 *Finite groups, abstract groups, Latin squares*

A group is called finite or infinite according to the number of elements in its set. In the case of finite groups this number is called the *order* of the group.

Exercise 7

1 Give the orders of the finite groups in Exercise 2.

2 Cayley tables for various groups of order 2 are shown below.

$+_2$	0	1
0	0	1
1	1	0

(i)

$\circ$	I	H
I	I	H
H	H	I

(ii)

$+$	E	O
E	E	O
O	O	E

(iii)

$\times_3$	1	2
1	1	2
2	2	1

(iv)

$\times$	P	N
P	P	N
N	N	P

(v)

$*$	e	a
e	e	a
a	a	e

(vi)

In (ii), H stands for a half turn about the origin; in (v), the table represents multiplication in the set of non-zero real numbers, with P, N denoting that the number is positive or negative; in (vi), the pattern common to all the tables is shown, using e for identity element, a for the other element and $*$ for the operation. This group is called the *abstract* group of order 2; it shows the pattern of all the others, abstracted from the particular elements and operations.

Make out a table for $\{e, a, b\}$, $*$ using the groups in Exercise 2, questions **4**, **5** and **6** to help you.

3 Try to invent a *different* table for $\{e, a, b\}$, $*$ still using e as identity element, and setting out the headings for the table in the same order. Remember that each element must have an inverse.

4 Notice that in question **3** you found that no element could appear twice in the same row. This can be explained as follows: Suppose $a*a = b$ and $a*b = b$; then $a*a = a*b$, and operating on the left by the inverse of a, $a = b$, which is not true. Why does this mean that each element has to appear at least once in each row? Is the same true for columns?

5 A square array of objects like the one shown at the side, in which each object appears exactly once in each row and exactly once in each column is called a Latin square.

p	q	r
q	r	p
r	p	q

All Cayley tables for groups are necessarily Latin squares. Try to complete in four different ways the four-by-four Latin square shown alongside.

	e	a	b	c
e	a	b	c	
a	.	.	.	
b	.	.	.	
c	.	.	.	

6 a Is the body of this table a Latin square?

 b Why is the system shown not a group? This example shows that for a system to be a group the Latin square property is not *sufficient*, although of course it is necessary.

$*$	e	a	b	c	d
e	e	a	b	c	d
a	a	b	c	d	e
b	b	e	d	a	c
c	c	d	a	e	b
d	d	c	e	b	a

7 Symmetry groups of geometrical figures

Much of the early deduction in the geometry course was based on the familiar fact that a rectangle fits its outline in four ways. This can be expressed in another way, namely that there are four transformations of the plane which leave the rectangle (as a whole) invariant. One is the identity transformation I; another is a half turn H in the plane about the centre O; the other two are reflections X and Y in the two axes of symmetry through O. We can complete a Cayley table for the set $\{I, H, X, Y\}$ and the operation of composition of transformations as indicated below.

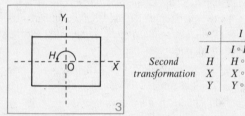

		First transformation			
$\circ$		I	H	X	Y
	I	$I \circ I$	$I \circ H$	$I \circ X$	$I \circ Y$
Second	H	$H \circ I$	$H \circ H$	$H \circ X$	$H \circ Y$
transformation	X	$X \circ I$	$X \circ H$	$X \circ X$	$X \circ Y$
	Y	$Y \circ I$	$Y \circ H$	$Y \circ X$	$Y \circ Y$

Exercise 8

1   Simplify the entries in the Cayley table shown above, and check that $\{I, H, X, Y\}, \circ$ has all the group properties.

Notice that composition of mappings is always associative since $(Q \circ P)(x)$ means $Q(P(x))$, so that

$$[R \circ (Q \circ P)](x) = R[(Q \circ P)(x)] = R[Q(P(x))], \text{ and}$$
$$[(R \circ Q) \circ P](x) = (R \circ Q)[P(x)] = R[Q(P(x))].$$

Copy and complete the *symmetry groups* of the following figures. In each case I is the identity transformation.

2   *Parallelogram*

$\circ$	I	H
I		
H		

H is a half turn about the centre.

3   *Rhombus*

$\circ$	I	H	A	B
I				
H				
A				
B				

H is a half turn about the centre. A and B are reflections in the two diagonals.

4 *Isosceles triangle*

$\circ$	I	M
I		
M		

M is a reflection in the axis of symmetry.

5 *Equilateral triangle.* The equilateral triangle XYZ fits its outline in six ways. The six transformations are:

I, as usual; two anticlockwise rotations in the plane about the centre O, through 120° and 240°, called R and R^2 respectively; and three reflections in the *fixed axes* AO, BO and CO, denoted by A, B and C.

Result of R Result of $A \circ R$ Result of B 4

From Figure 4 we can see that $A \circ R = B$. Check that $R \circ A = C$.

Copy and complete the Cayley table, taking great care about the order. Since $R \circ A \neq A \circ R$, for example, the table will not be symmetrical about the main diagonal. Towards the end you may find the Latin square property helpful.

$\circ$	I	R	R^2	A	B	C
I	I	R	R^2	A	B	C
R	R					C
R^2	R^2					
A	A	B				
B	B					
C	C					

6 *Square.* The square fits its outline in eight ways; it has the symmetries of the rectangle and four more besides. These are: rotations in the plane about O of 90° and 270°, called R and R^3; and two reflections M_1 and M_2 in the diagonals which are regarded as *fixed.* Note that R^2 in this table is a half turn about O.

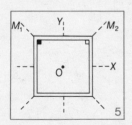

$\circ$	I	R	R^2	R^3	X	M_1	Y	M_2
I								
R								
R^2								
R^3								
X								
M_1								
Y								
M_2								

Make a neat sketch to show each of the eight positions of the square which result from the above transformations. Then use the sketches to help you to complete the Cayley table.

7 Some of the groups you have met are commutative, or *abelian* (named after the Norwegian mathematician Abel). Others are not, e.g. the last two. By looking for symmetry about the main diagonals in the Cayley tables, decide which of the groups you have met so far are abelian.

8 Isomorphism

The composition tables for the addition of even and odd numbers and for the symmetry group of a parallelogram, both of order 2, are shown alongside. Note that:

+	E	O
E	E	O
O	O	E

∘	I	H
I	I	H
H	H	I

If E corresponds to I and O corresponds to H, then:

$$E+E \leftrightarrow I \circ I, \quad E+O \leftrightarrow I \circ H,$$
$$O+E \leftrightarrow H \circ I, \quad O+O \leftrightarrow H \circ H.$$

There is a *one-to-one correspondence* between the sets $\{E, O\}$ and $\{I, H\}$ which *preserves the operations*. The groups $\{E, O\}$, + and $\{I, H\}$, ∘ are said to be *isomorphic* to each other.

In general, there is an *isomorphism* between two groups $\{a, b, c, ...\}$, ∗ and $\{a', b', c', ...\}$, ∘ if there is a one-to-one correspondence between the elements of the sets which preserves the group operations, i.e.

if $a \leftrightarrow a'$ and $b \leftrightarrow b'$, then $a * b \leftrightarrow a' \circ b'$.

An isomorphism you have already met is that between the (infinite) group of positive real numbers under multiplication and the (infinite) group of real numbers under addition, i.e. between the groups $\{m, m', m'', ...\}$, × and $\{a, a', a'', ...\}$, +.

If $m \leftrightarrow a$ and $m' \leftrightarrow a'$, then $m \times m' \leftrightarrow a+a'$, so that multiplication in the first group can be 'converted' into addition in the second group.

The logarithmic function (to any base b) is a suitable mapping for the isomorphism $l : m \rightarrow a$.

If $\log_b m = a$, and $\log_b m' = a'$, then $\log_b(m \times m') = \log_b m + \log_b m' = a+a$, i.e. $\log_b$ maps $m \times m'$ to $a+a'$.

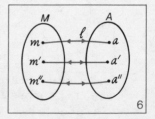

6

Thus the property $\log_b (m \times m') = \log_b m + \log_b m'$ ensures that the function $\log_b$ $(b > 0, b \neq 1)$ sets up the above isomorphism.

We made use of this and related results in the chapter on Logarithms in Book 5.

It is worth noting that if two groups are isomorphic, the properties of one may be restated in terms of the properties of the other.

Exercise 9

1 List the pairs of corresponding elements in an isomorphism between the groups in questions *4* and *5* of Exercise 2.

2 List a correspondence between the isomorphic groups in questions *9* and *12* of Exercise 2.

3 *a* Make a Cayley table for $\{1, 3, 5, 7\}$, $\times_8$.
  b Is this group isomorphic to the group in question *11* of Exercise 2?
  c Is it isomorphic to the group in question *9* of Exercise 2? If not, why not?

4 Is the group in question *8* in Exercise 2 isomorphic to the group in question *9*? Try writing the first table in the order 1, 2, 4, 3. List the correspondence now.

5 *a* Regard each of the four Latin squares in question *5* of Exercise 7 as the body of a group table. Which group is isomorphic to $\{1, 3, 5, 7\}$, $\times_8$?
  b Set that group aside, and consider the other three. Rearrange the order of the elements in two of them so as to show that *each* of the three is isomorphic to $\{0, 1, 2, 3\}$, $+_4$.

  We have seen that the names of the elements in an abstract group are not important; hence there are just two abstract groups of order 4 (called the Klein four-group and the cyclic group of order 4).

6 Is $\{1, 2, 3, 4, 5, 6\}$, $\times_7$ isomorphic to the symmetry group of the equilateral triangle? Is it commutative?

7 Why cannot two finite groups of different orders be isomorphic? Is Z, $+$ isomorphic to $\{\ldots -4, -2, 0, 2, 4, \ldots\}$, $+$?

8 Make a Cayley table for the group formed by the following six functions under the operation 'composition of functions'. Then comment.

$$I : x \to x \qquad R : x \to 1 - \frac{1}{x} \qquad R : x \to \frac{1}{1-x}$$

$$A : x \to \frac{1}{x} \qquad B : x \to \frac{x}{x-1} \qquad C : x \to 1 - x$$

Topics to Explore

(i) For further reading see *The Fascination of Groups*, by F. J. Budden (C.U.P.).

(ii) Use library resources to find out more about groups, rings and fields, their history and their modern uses, etc.

Summary

1   A *binary operation* combines two elements to produce a unique single element.

2   *Composition tables* are useful for displaying the elements produced by a binary operation on the members of a given set.

3   A *group* is a system consisting of a *set of elements* and an *operation*, with the following properties:

    (i) *Closure.* The operation always results in an element of the set.
    (ii) *Associativity.* The operation is associative.
    (iii) *Identity element.* The set contains an identity element.
    (iv) *Inverse.* Each element in the set has an inverse element in the set.

4   A *group* may be *finite* or *infinite*. The order of a finite group is the number of elements it contains.

     A group may, or may not, be *commutative*. If it is, its composition table is symmetrical about the main diagonal.

5   For each symmetry of a plane figure there is a *transformation of the plane* which maps the figure onto itself. These transformations form a group, called the *symmetry group* of the figure.

6   Two groups $\{a, b, c, \ldots\}$, $*$ and $\{a', b', c', \ldots\}$, $\circ$ are *isomorphic* if there is a one-to-one correspondence between the elements of the sets which preserves the group operations, i.e.

$$\text{if } a \leftrightarrow a' \text{ and } b \leftrightarrow b', \text{ then } a * b \leftrightarrow a' \circ b'.$$

Revision Exercises

Revision Exercise on Chapter 1
Quadratic Equations and Functions

Revision Exercise 1

1 Solve the following quadratic equations, $x \in R$:

   a $6x^2 + x - 1 = 0$ b $x^2 - 4x - 2 = 0$ c $2x^2 - 3x + 4 = 0$

2 Find h, given that $\frac{1}{2}$ is a root of $4x^2 + 2hx - 7 = 0$.

3 State the nature of the roots of the equations:

   a $3x^2 - 7x + 2 = 0$ b $2x^2 + 9x + 3 = 0$ c $2x^2 + x + 5 = 0$

4 In each of the following, find k so that:
   (1) the roots are equal   (2) the roots are real and unequal.

   a $x^2 - 8x + k = 0$ b $x^2 - (k+1)x + k = 0$

   c $x^2 + (k-3)x + k = 0$ d $kx^2 + 3x = k - 5$

5 Find m for the line $y = mx$ to

   a be a tangent to the circle $x^2 + y^2 - 8x - 4y + 16 = 0$.
   b cut the circle in two distinct points.
   Illustrate your answers with a diagram.

6 Find the equations of the two tangents to the circle $x^2 + y^2 = 5$ which have gradient 2. What are the coordinates of the points of contact? Illustrate by a sketch.

7 Find the equations of the tangents from the point $(0, 4)$ to the circle $x^2 + y^2 = 4$. Illustrate by a sketch.

8 Sketch the curves:

   a $y = x^2 - 4x$ b $y = -x^2 - 6x + 7$ c $y = 2x^2 + 4x + 5$

9 For the quadratic functions f defined by:

  a $f(x) = 3x^2 - 16x + 5$
  b $f(x) = 1 + 3x - 4x^2$,
    find x for which f is (1) zero (2) positive (3) negative.

10 If α and β are the roots of the quadratic equation $x^2 + rx + s = 0$, express each of the following in terms of r and s:

  a $\alpha^2 + \beta^2$ *b* $\dfrac{1}{\alpha} + \dfrac{1}{\beta}$ *c* $\dfrac{\alpha}{\beta} + \dfrac{\beta}{\alpha}$ *d* $(\alpha - \beta)^2$

11 Find the simplest quadratic equation with roots:

  a $7, -5$ *b* $\frac{2}{3}, \frac{3}{4}$ *c* $2 \pm \sqrt{3}$ *d* $m, \dfrac{1}{m} (m \neq 0)$

12 Given the quadratic equation $x^2 + px + q = 0$, what can you deduce about p and q if:

  a one root is the additive inverse of the other
  b one root is the multiplicative inverse of the other?

13 Use the discriminant to show that the roots of the equation $12x^2 - 7x - 12 = 0$ are real, rational and distinct. Solve the equation.

14 Which of the following are true and which are false for $x \in R$? $x^2 - 6x + 12$ cannot be:

  a zero   *b* positive   *c* negative

  d greater than 3   *e* less than 3.

15a If $k = \dfrac{x^2 + 2x + 3}{x + 1}$, form a quadratic equation in x.

  b Show that for real roots, $k \geqslant 2\sqrt{2}$ or $k \leqslant -2\sqrt{2}$.

16 The height h metres an object will rise above the ground after t seconds when projected with an upward velocity of 30 m/s is given by $h = 30t - 5t^2$. Find:

  a when it reaches the ground
  b when it is 25 m high
  c if it can reach a height of 50 m.

Revision Exercise on Chapter 2
Polynomials, the Remainder Theorem and Applications

Revision Exercise 2

1 Arrange the following polynomials in descending powers of x, and state the degree of each:

  a $x^2 - 3x + x^3 - 3$ b $x(3 - 2x)^2$ c $x(x + 1)(1 - x^2)$

2 Find the values of p and q in each of the following:

  a $px^2 + qx - 7 = (2x + 1)(x - 7)$
  b $(px + 2)(x - q) = 3x^2 - 10x - 8$

3 $f(x) = x^3 - 4x^2 - 3x + 1$.

  a Evaluate $f(0), f(1), f(-1)$ and $f(4)$.
  b Find $f'(x)$, and evaluate $f'(3)$ and $f'(-\frac{1}{3})$.

4 Calculate the value of $2x^3 - 8x + 6$ when $x = \frac{1}{2}$ and when $x = 0.1$.

5 Find the quotient and remainder when $x^2 - 8x + 11$ is divided by $x - 3$; express your result in the form $f(x) = (x - h)Q(x) + R$.

6 Repeat question 5 when

  a $x^3 + 7x^2 - 3x + 9$ is divided by $x + 2$
  b $4x^3 - 39x + 11$ is divided by $x - 3$

7 Find the remainder on dividing:

  a $x^3 + x^2 - 6x + 7$ by $x + 1$ b $x^4 - 5x^2 + 11$ by $x - 4$
  c $x^7 + x^5 + x^3 + x$ by $x - 1$

8 Find the remainder on dividing:

  a $4x^2 + 2x - 2$ by $2x - 1$ b $4x^2 - 8x + 2$ by $2x - 3$

  c $6x^3 - 2x^2 - x + 7$ by $3x + 2$

9 Show that $x - 2$ is a factor of $x^3 + 2x^2 - 5x - 6$, and find the other factors.

10   Factorise:

a $x^3 - 4x^2 - x + 4$

b $y^4 - 5y^2 - 36$

c $6z^3 + 19z^2 + z - 6$

d $x^3 + x^2 + x - 3$

11   Find k if $2x^3 + x^2 + kx - 8$ is divisible by $x + 2$.

12   Find a and b if $x - 1$ and $2x + 1$ are factors of $6x^3 - 7x^2 + ax + b$.

13   The same remainder is obtained when $x^2 + 3x - 2$ and $x^3 - 4x^2 + 5x + p$ are divided by $x + 1$. Find p.

14   If $2x^3 - 3x^2 + ax + b$ is divisible by $x + 1$ show that $b = a + 5$.

15   Solve the following equations, $x \in R$:

a $3x^3 - 11x^2 + 5x + 3 = 0$

b $x^3 - 19x + 30 = 0$

c $2x^2(x + 2) = (x + 3)(x + 2)$

d $x^4 = 3x^2 + 2x$

16   Find the zeros of the polynomial $f(x) = x^4 - x^2 + 4x - 4$.

17a   Show that $2x + 3$ is a factor of $f(x) = 6x^3 - 29x^2 - 85x - 42$, and hence factorise the polynomial completely.

b   Write down the roots of the equation $f(x) = 0$.

18   Find the coordinates of the points at which the graph of the function $f : x \rightarrow 4x^3 - 13x - 6$ meets the x-axis.

19   Verify that $0 \cdot 6$ is an approximate root of $10x^3 - 3x^2 - 1 = 0$.

20   Show that one root of the equation $x^3 - x^2 - 5x + 2 = 0$ lies between 0 and 1, and another between 2 and 3. Find the smaller of these roots to one decimal place.

21a   Show that the curve $y = 3x^4 - 8x^3 - 6x^2 + 24x$ cuts the x-axis at the origin and at another point whose x-coordinate lies between -1 and -2.

b   Find the stationary points of the curve, and determine the nature of each. Sketch the curve.

Revision Exercise on Chapter 3
The Exponential and Logarithmic Functions

Revision Exercise 3

1   If $a = 4$, $b = 8$ and $c = 16$, find the values of:

   a $a^{1/2} + b^0 + c^{-1/2}$ b $a^{-1} + b^{-1} + c^{-1}$ c $a^{3/2} b^{-2/3} c^{3/4}$

2   Simplify:

   a $x^{1/2}(x^{1/2} - x^{-1/2})$ b $(x^{1/2} + 1)(x^{1/2} - 1)$ c $(x^{1/2} - x^{3/2})^2$

3   Which of the following are true and which are false?

   a $(8^3)^{-2} > (8^2)^{-3}$ b $3^{-2} > 2^{-3}$ c $2^a > 2^b \Rightarrow a > b$

4   Given $p = x^{1/2} - 3$, $q = (x^{1/2} - 1)(x^{1/2} + 2)$ and
   $r = (x^{1/2} + 1)(2x^{1/2} - 1)$, show that $p + r = 2q$.

5 a Sketch the graphs of $f : x \rightarrow 4^x$ and $g : x \rightarrow 4^{-x}$ for $\{x : -2 \leqslant x \leqslant 2,$
      $x \in R\}$.
  b Write down the coordinates of the point of intersection of the graphs.
  c Give the equation of the axis of symmetry in the diagram.
  d Calculate the gradient of the line joining the points on the graph of
      f with x-coordinates 0 and 2.

6   f and f^{-1} denote inverse functions. Fill in the blank in the following:
   If $f : x \rightarrow 8^x$, then $f^{-1} : x \rightarrow \ldots$.

7   Express the following in equivalent logarithmic form:

   a $16 = 2^4$ b $1 = 3^0$ c $2 = 4^{1/2}$ d $\frac{1}{100} = 10^{-2}$

8   Express the following in equivalent index form:

   a $\log_9 81 = 2$ b $\log_4 16 = 2$ c $\log_{16} 4 = \frac{1}{2}$ d $\log_2 \frac{1}{4} = -2$

9   Find a in each of the following:

   a $\log_a 25 = 2$ b $\log_a 5 = \frac{1}{2}$ c $\log_a 16 = 4$ d $\log_4 4 = a$

   e $\log_8 1 = a$ f $\log_{1/4} 64 = a$ g $\log_2 a = 3$ h $\log_{10} a = 1$

10   Show that if $\log_2 x = \pi$, then $8 < x < 16$.

11a On 2-mm squared paper draw the graphs of $y = \log_{10} x$ and $y = \log_{1/10} x$, taking $x = \frac{1}{10}$, 1, 10, 50 and 100, and scales of 10 units to 2 cm horizontally and 1 unit to 2 cm vertically.

b Write down the coordinates of the point of intersection of the graphs.

c Give the equation of the axis of symmetry in the diagram.

d Estimate the y-coordinates of the points on the graphs for which $x = 30$.

12 Simplify:

a $\log_{10} 25 + \log_{10} 8 - \log_{10} 2$ *b* $\log_2 \frac{25}{24} + \log_2 \frac{9}{40} - \log_2 \frac{15}{4}$

13 Given $\log_a 128 - \log_a 64 + \log_a \frac{1}{8} = 2$, find a.

14 Solve, $x \in R$:

a $\log_2 x + \log_2 (x - 2) = 3$ *b* $\log_5 (3 - 2x) + \log_5 (2 + x) = 1$

15 Simplify:

a $2 \log_9 2 + 3 \log_9 3 - \log_9 36$ *b* $2 \log_a a + 3 \log_a a^2 + 4 \log_a \dfrac{1}{a}$

16 Simplify:

a $\log 2 + \log \sqrt{2}$ *b* $\log 2 - \log \sqrt{2}$ *c* $\log 2 / \log \sqrt{2}$

17 If $\log y = \log 2 + 3 \log x$, express y in terms of x.

18 Find x to 2 significant figures if:

a $2^x = 5$ *b* $3^{4x} = 4$ *c* $x = 3 \cdot 8^{0 \cdot 12}$

19 Find the least integer n for which $\dfrac{1}{2} + \dfrac{1}{2^2} + \dfrac{1}{2^3} + \ldots + \dfrac{1}{2^n}$ exceeds $\dfrac{99}{100}$.

20a Find the sum of n terms and the sum to infinity of the geometric series $1 + \frac{1}{2} + \frac{1}{4} + \frac{1}{8} + \frac{1}{16} + \ldots$.

b Find the least value of n for which these sums differ by less than 0·001.

21 Taking $e = 2 \cdot 72$, calculate to 2 significant figures:

a $\log_e 4$ *b* $\log_e 6$ *c* $\log_\pi e$ *d* e^π

22 If $\dfrac{\log_2 a}{\log_2 b} = m$ and $\dfrac{\log_3 a}{\log_3 b} = n$, find a relation between m and n.

23   Under certain conditions the pressure P of a given mass of gas varies inversely as V^n, where V is the volume of the gas and n is a constant. When $P = 3$, $V = 10$ and when $P = 8$, $V = 5$. Find n, and give the formula for P in terms of V.

24   Corresponding readings of x and y in an experiment are given in the table

x	1·0	1·5	2·0	3·0	4·0
y	2·50	8·42	20·0	67·5	160

Show that $y = kx^n$, and find k and n.

25   The mass of a radioactive element decreases at a rate given by $m_t = m_0 e^{-0.01t}$, where t is the time in years. Find:

a   the mass of 250 mg of the element after a century
b   the half-life of the element.

Revision Exercise on Chapter 4
Pattern and Structure: Groups

Revision Exercise 4

1   State whether each of the following is a unary or a binary operation, and give examples of sets of elements on which it might operate:

a   doubling   b   division

c   matrix multiplication   d   complementation

2   Here is an operation table for the set $S = \{1, 3, 5, 6, 10, 15, 30\}$ under the binary operation $*$ defined by '$a * b$ is the least common multiple of a and b'.

$*$	1	3	5	6	10	15	30
1	1	3	5	6	10	15	30
3	3	3	15	6	30	15	30
5	5	15	5	30	10	15	30
6	6	6	30	6	30	30	30
10	10	30	10	30	10	30	30
15	15	15	15	30	30	15	30
30	30	30	30	30	30	30	30

a   Test the following properties: closure, associativity, identity element, inverses. Also test for commutativity.

b   Can you solve the equation $x * a = b$ for all $a, b \in S$ and x a variable on S? If not, can you find a condition on a and b which ensures that $x * a = b$ has a solution?

3   For each of these systems the operation is closed and associative, and it is also commutative. Check the other two group properties.

a $N, +$; $W, +$; $Z, +$; $Q, +$; $R, +$.
b $N, \times$; $W, \times$; $Z, \times$; $Q, \times$; $Q^*, \times$; $R, \times$; $R^*, \times$,
    where Q^*, R^* are the sets of all non-zero rational and real numbers respectively.

4   A system $A, +, \times$ is called a *field* if

(i) $A, +$ is a commutative group
(ii) $A^*, \times$ is a commutative group (A^* is the set A without the identity element of $A, +$)
(iii) $\times$ is distributive over $+$.

Use the results of question 3 to decide which of these is a field:

a $N, +, \times$ b $Z, +, \times$ c $Q, +, \times$ d $R, +, \times$

5   A set $\mathscr{P}(E)$ consists of all subsets of the set E. Consider the binary operation intersection ($\cap$) on $\mathscr{P}(E)$.

a   Test for the four group properties, and also for commutativity.
b   Does the equation $X \cap A = B$, for $A, B \in \mathscr{P}(E)$ and X a variable on $\mathscr{P}(E)$, always have a solution?
c   If $B \subset A$, does $X \cap A = B$ have a solution?

6   Mappings f_1, f_2, f_3 and f_4 from R^* to R^* (see question 3) are defined by the formulae $f_1(x) = x, f_2(x) = -x, f_3(x) = \dfrac{1}{x}$ and $f_4(x) = -\dfrac{1}{x}$.

a   Make a Cayley table for the set $\{f_1, f_2, f_3, f_4\}$ and the operation of composition of mappings.
b   Explain why the system forms a group. Is the group commutative?

7   Which of the following equations can be solved by using an appropriate inverse? Find the solution in each case. One of the other equations does in fact have solutions; find the solution set of this equation.

a $3x = 7$; x a variable on Z.

b $0 \times x = 3$; x a variable on R.

c $X + \begin{pmatrix} 1 & 1 \\ 1 & 1 \end{pmatrix} = \begin{pmatrix} 1 & 2 \\ 2 & 1 \end{pmatrix}$; X a variable on the set of all 2×2 matrices.

d $X \times \begin{pmatrix} 1 & 1 \\ 1 & 1 \end{pmatrix} = \begin{pmatrix} 1 & 2 \\ 2 & 1 \end{pmatrix}$; X a variable on the set of all 2×2 matrices.

e $X \cap \{3,4,5\} = \{3,4\}$; X a variable on the set of subsets of $\{1,2,3,...,9\}$.

8 Each of these letters has one or more symmetries. Give the *order* of the group of symmetries for each:

a H   **b** A   **c** Z   **d** E   **e** O

9 The shapes in Figure 1 have symmetry groups of orders 1 and 2 respectively.

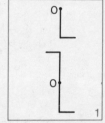

a Draw shapes consisting of identical letters L drawn from O, which have symmetry groups of orders 3 and 4.

b Describe such a figure with a symmetry group of order n $(n \in N)$.

10 In each of the following, two groups are given; either specify an isomorphism between them, or else give a reason to show that they are not isomorphic.

a $\{1,2\}, \times_3$; $\{0,1\}, +_2$ **b** $\{0,1,2\}, +_3$; $\{0,1\}, +_2$

c {even integers}, $+$; {all integers}, $+$

d $\{1,3,7,9\}, \times_{10}$; $\{I, R, R^2, R^3\}, \circ$ where R is a rotation about O of $90°$.

e $\{\log_b x : x > 0, x \in R\}, +$; $\{\log_a x : x > 0, x \in R\}, +$ where a and b are fixed positive real numbers not equal to unity.

f Symmetry group of the equilateral triangle; $\{I, R, R^2, R^3, R^4, R^5\}, \circ$ where R is a rotation about O of $\frac{1}{3}\pi$ radians.

Cumulative
Revision
Section

Cumulative Revision Section (Books 8-9)

Revision Topic 1. Systems of Equations
Exercise 1

(*All the variables in this Exercise are on R.*)

1   The equations of the lines forming the sides of a triangle are $6x + y = 16, 5x - 6y = -14$ and $x + 7y = -11$. Find the coordinates of the vertices of the triangle.

Solve the systems of equations in questions **2–4**:

2   $x + y + z = 1$
    $x + y - z = -1$
    $x - y + z = 3$

3   $x + 2y + 3z = -1$
    $2x + y - z = 5$
    $3x - y + 2z = 4$

4   $2u + 3v + 4w = 10$
    $4u + 3v + 2w = 8$
    $3u - v - 3w = -4$

5   A function f is defined by $f(x) = ax^2 + bx + c$. Given $f(2) = 7$, $f(0) = -3$ and $f(-1) = -2$, find a, b and c.

6   The circle $x^2 + y^2 + 2gx + 2fy + c = 0$ passes through the points $(2, 1)$, $(2, -3)$ and $(0, -1)$. Find the equation of the circle.

Solve the systems of equations in questions **7–9**:

7   $x - 2y - 1 = 0$
    $3x^2 - 4y^2 + 1 = 0$

8   $2x + 3y = 4$
    $x^2 + y^2 = 5$

9   $3x - 4y = 6$
    $x^2 - xy + y^2 = 4$

10 Find the points of intersection of the line $y = 2x + 2$ and the parabola $y = x^2 + 2x + 1$. Illustrate by a sketch.

11 Find the points of intersection of the circle, centre O, radius 2, and the line through the points $(4, 4)$ and $(0, -4)$. Illustrate by a sketch.

12 The line with gradient $\frac{3}{4}$ passing through the centre of the circle $x^2 + y^2 - 4x - 10y + 4 = 0$ cuts the circle at A and B. Find the coordinates of A and B, and illustrate by a sketch.

Revision Topic 2. Sequences and Series

Reminders

1   *Sequences.* The terms of a sequence may be defined by a formula, e.g. $u_n = 2n - 1$ gives the sequence 1, 3, 5, ...

2   In an *arithmetic sequence*, $u_2 - u_1 = u_3 - u_2 = \ldots = u_n - u_{n-1}$; that is, it has a *common difference*.

   The nth term of the arithmetic sequence $a, a+d, a+2d, \ldots$ is

$$u_n = a + (n-1)d.$$

3   *Arithmetic series.* The standard series is $a + (a+d) + (a+2d) + \ldots + [a + (n-1)d]$.

   The sum of n terms, $S_n = \frac{1}{2}n[2a + (n-1)d]$.

4   In a *geometric sequence*, $\dfrac{u_2}{u_1} = \dfrac{u_3}{u_2} = \ldots = \dfrac{u_n}{u_{n-1}}$; that is, it has a *common ratio.*

   The nth term of the geometric sequence $a, ar, ar^2, \ldots$ is $u_n = ar^{n-1}$.

5   *Geometric series.* The standard series is $a + ar + ar^2 + \ldots + ar^{n-1}$.

   The sum of n terms, $S_n = \dfrac{a(1-r^n)}{1-r}$, or $\dfrac{a(r^n-1)}{r-1}. (r \neq 1)$.

   The sum to infinity, $S_\infty = \dfrac{a}{1-r}$, provided that $-1 < r < 1$.

Exercise 2

1   The nth term of a sequence is $u_n = \frac{1}{2}(n-2)(2n+1)$. Find the twelfth term, and the value of n when $u_n = 26$.

2   The first term of an arithmetic series is 18, and the fourth term is 9. Show that the sum of the first 4 terms is equal to the sum of the first 9 terms.

3   For a certain series the sum of n terms is $S_n = \frac{1}{2}n(n-1)(n+3)$. Calculate S_1, S_2 and S_3. Hence find the first three terms of the series.

4   Calculate the sum to infinity of the geometric series which has first term 120 and fourth term -15.

5   The third term of a geometric series is $\frac{4}{3}$ and the sixth term is $\frac{32}{81}$. Find the difference between the sum of the first five terms and the sum to infinity of the series.

6 *a* Find the sum of n terms and the sum to infinity of the geometric series whose nth term is $1/3^n$.

   b Find the smallest value of n for which the sum of n terms and the sum to infinity differ by less than 0·001.

7 *a* Determine which one of the following is an arithmetic sequence and which one is a geometric sequence:

   $(1)\sqrt{3}, 3, 3\sqrt{3}, 9, \ldots$ $(2)\sqrt{3}, \sqrt{4}, \sqrt{5}, \sqrt{6}, \ldots$ $(3)\sqrt{3}, 2\sqrt{3}, 3\sqrt{3}, 4\sqrt{3}, \ldots$

   b For each of the two sequences write down the fifth term, and find a formula for the nth term.

   c Find also a formula for the sum of n terms of the corresponding series.

8 *a* Write down the sum to infinity (S_1) of the geometric series $1+q+q^2+q^3+\ldots$, and state the necessary restriction on q.

   b Write down the sum to infinity (S_2) of the series formed from the squares of the terms of the first series, and show that $S_1 : S_2 = (1+q):1$.

9 *a* Show that the sequence formed by the differences between successive pairs of terms in each of the following sequences is an arithmetic sequence:

   $(1)1, 4, 9, 16, 25, \ldots$ $(2)1 \times 2, 2 \times 3, 3 \times 4, 4 \times 5, \ldots$

   b In each case find a formula for the nth term of the given sequence and of the arithmetic sequence.

   c Give the hundredth term of each of the four sequences.

10   An arithmetic series has first term a and common difference d. In the usual notation, write down formulae for S_n, S_{2n} and S_{3n}. Show that $S_{2n} - S_n = \frac{1}{3}S_{3n}$.

11   £100 is invested at 8% per annum compound interest. Find the amount after 1, 2 and 3 years, to the nearest £.
   State the growth factor, and calculate the amount after 10 years.

Revision Topic 3. Matrices

Reminders

1   A *matrix* is a rectangular array of members arranged in rows and columns. With m rows and n columns the order of the matrix is $m \times n$.

2 a $\begin{pmatrix} a & b \\ c & d \end{pmatrix} \pm \begin{pmatrix} p & q \\ r & s \end{pmatrix} = \begin{pmatrix} a \pm p & b \pm q \\ c \pm r & d \pm s \end{pmatrix}$

   b $I = \begin{pmatrix} 1 & 0 \\ 0 & 1 \end{pmatrix}$ is an identity matrix.

   c $k \begin{pmatrix} a & b \\ c & d \end{pmatrix} = \begin{pmatrix} ka & kb \\ kc & kd \end{pmatrix}$

   d $\begin{pmatrix} a & b \\ c & d \end{pmatrix} \begin{pmatrix} p & q \\ r & s \end{pmatrix} = \begin{pmatrix} ap+br & aq+bs \\ cp+dr & cq+ds \end{pmatrix}$

3   The *inverse* of $A = \begin{pmatrix} a & b \\ c & d \end{pmatrix}$ is $A^{-1} = \dfrac{1}{ad-bc} \begin{pmatrix} d & -b \\ -c & a \end{pmatrix}$, $ad - bc \neq 0$.

   $ad - bc$ is the *determinant* of matrix A. $A^{-1}A = AA^{-1} = I$.

4   2×2 matrices can be associated with transformations of the plane, and products of such matrices are then associated with compositions of transformations.

Exercise 3

1   $A = \begin{pmatrix} 2 & 3 \\ -1 & 0 \end{pmatrix}$, $B = \begin{pmatrix} 0 & -3 \\ 1 & -1 \end{pmatrix}$ and $C = \begin{pmatrix} -2 & 0 \\ 2 & 1 \end{pmatrix}$. Find in simplest form:

   a $A + B + C$ b $A - B - C$ c $2A + 3B - 4C$

2   Given $3 \begin{pmatrix} p & -2 \\ 0 & q \end{pmatrix} - 2 \begin{pmatrix} -6 & r \\ s & 2 \end{pmatrix} = \begin{pmatrix} 6 & 0 \\ -1 & -1 \end{pmatrix}$, find p, q, r and s.

3   Express in simplest form:

   a $\begin{pmatrix} \sqrt{2} & \sqrt{2} \\ \sqrt{2} & \sqrt{2} \end{pmatrix} \begin{pmatrix} \sqrt{2} & \sqrt{2} \\ \sqrt{2} & -\sqrt{2} \end{pmatrix}$ b $\begin{pmatrix} \cos\theta & \sin\theta \\ \cos\theta & \sin\theta \end{pmatrix} \begin{pmatrix} \cos\theta & -\sin\theta \\ \sin\theta & \cos\theta \end{pmatrix}$

4   $A = \begin{pmatrix} 3 \\ 2 \end{pmatrix}$, $B = (-1 \quad 4)$ and $C = \begin{pmatrix} 1 & -2 \\ -3 & 0 \end{pmatrix}$. Evaluate all the products of two of these matrices that exist.

5    $P = \begin{pmatrix} 2 & 5 \\ 3 & -4 \end{pmatrix}$, $Q = \begin{pmatrix} x \\ y \end{pmatrix}$ and $R = \begin{pmatrix} -4 \\ -6 \end{pmatrix}$. If $PQ = R$, find x and y.

6    $M = \begin{pmatrix} 2 & 1 \\ 1 & 2 \end{pmatrix}$, $N = \begin{pmatrix} 3 & 0 \\ 0 & 1 \end{pmatrix}$ and $X = \begin{pmatrix} a & b \\ c & d \end{pmatrix}$. If $MX = XN$ show
      that $c = a$ and $d = -b$.

7   Under a certain mapping, $P(x, y) \to P'(x', y')$ such that $x' = x - y$
      and $y' = 2x + 3y$. Write down the matrix associated with the
      mapping, and hence find the images of the points $(2, 1)$, $(4, 4)$ and
      $(-3, -2)$.

8    $M = \begin{pmatrix} \frac{1}{2}\sqrt{3} & -\frac{1}{2} \\ \frac{1}{2} & \frac{1}{2}\sqrt{3} \end{pmatrix}$. Prove that $M^3 = \begin{pmatrix} 0 & -1 \\ 1 & 0 \end{pmatrix}$, and hence state
      the transformation associated with M.

9    $A = \begin{pmatrix} 5 & 2 \\ 7 & 3 \end{pmatrix}$. Write down the inverse matrix A^{-1}. Show that the
      squares of A and A^{-1} are also inverses of one another.

10   $N = \begin{pmatrix} \frac{1}{\sqrt{2}} & -\frac{1}{\sqrt{2}} \\ \frac{1}{2} & \frac{1}{2} \end{pmatrix}$. Write down the inverse matrix N^{-1}. Hence or
      otherwise solve the system of equations $x' = \frac{1}{\sqrt{2}}x - \frac{1}{\sqrt{2}}y$, $y' = \frac{1}{2}x + \frac{1}{2}y$
      for x and y, in terms of x' and y'.

11   Figure 1 shows the routes connecting towns a_1, a_2, b_1, b_2, b_3, c_1
      and c_2; the distances in kilometres are marked in colour.

   a Copy and complete the tables giving the number of direct links
      between the towns.

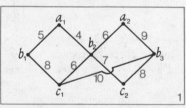

	b_1	b_2	b_3
a_1			
a_2			

	c_1	c_2
b_1		
b_2		
b_3		

   b Express the entries in the tables in matrix form, and deduce a matrix
      which gives the number of routes from a_1 and a_2 to c_1 and c_2 via
      b_1, b_2 and b_3 only.

   c Write down a matrix which shows the shortest distances from a_1
      and a_2 to c_1 and c_2.

12   Show that the product of the matrix $\begin{pmatrix} \sin \alpha & -\cos \alpha \\ \cos \alpha & \sin \alpha \end{pmatrix}$ and its
      transpose is the 2×2 identity matrix.

Revision Topic 4. Functions

Reminders

1   A *function*, or *mapping*, from a set A to a set B is a special kind of relation in which each element of A is related to exactly one element of B.

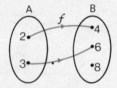 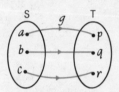

$f:A \to B; f(x) = 2x,$ or $x \to 2x$
Domain $\{2, 3\}$, *range* $\{4, 6\}$

S and T are in
one-to-one correspondence

2   *Composition of functions.* If $f:A \to B$ and $g:B \to C$, then the composite function $g \circ f:A \to C$ is defined by the formula $(g \circ f)(x) = g(f(x))$, $x \in A$. $g \circ f$ means f first, then g.

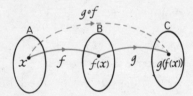 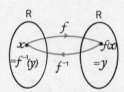

3   *Inverse functions.* A function $f:A \to B$ has an inverse function $g:B \to A$ if the sets are in one-to-one correspondence.
When g exists, it is denoted by f^{-1}. $f^{-1}(y) = x \Leftrightarrow f(x) = y.$

Exercise 4

1   Let $f:R \to R$ be defined by $f(x) = x^2 + x + 1$. Find $f(h)$, $f(4+h)$ and $f(4+h) - f(h)$ in simplest form.

2   Function $f:R \to R$, defined by $f(x) = ax^2 + bx$ is such that $f(2) = 4$ and $f(-3) = -21$. Find the constants a and b, and hence calculate $f(5)$.

3 Function $g:R \to R$ is defined by $g(x) = px^2 + qx + r$, where p, q and r are constants.

  a Given $g(0) = -6$, $g(1) = 5$, $g(-2) = 8$, find p, q and r.
  b Find the zeros of g.
  c Show that $g(\sqrt{2}-1) = 7(1-\sqrt{2})$.

4 $f:x \to 3x+2$ and $g:x \to x^2+1$ are functions from R to R. Find x in each of the following

  a $(f \circ g)(x) = 80$ b $(g \circ f)(x) = 50$ c $(f \circ g)(x) = (g \circ f)(x)$

5 Functions $f:R \to R$ and $g:R \to R$ are defined by $f(x) = x+3$ and $g(x) = 2x^2 - x + 1$.

  a Find $(g \circ f)(x)$ and $(f \circ g)(x)$. Hence find a such that $(g \circ f)(a) = (f \circ g)a$.
  b Find formulae for $(f \circ f)(x)$ and $(f \circ f \circ g)(x)$.

6 $h:R \to R$ and $k:R \to R$ are functions defined by $h(x) = x^2$ and $k(x) = x+1$. Express each of the following functions on R as a composition involving h or k or both.

  a $f:x \to (x+1)^2$ b $s:x \to x^4$ c $t:x \to x^2+1$

  d $u:x \to x+2$ e $v:x \to (x+2)^2$ f $w:x \to x^4+1$

7 $A = \{$non-negative real numbers$\}$ and function $g:A \to R$ is defined by $g(x) = 9 - 4x^2$. Sketch the graph of g, and state the range of g.
    Does g have an inverse function g^{-1}? If so, explain why, and find a formula for g^{-1}.

8 In each of the following find a formula for f^{-1}, the inverse of f. State a suitable domain for f in each case.

  a $f(x) = x^3 + 1$ b $f(x) = \frac{1}{2}(2x-3)$ c $f(x) = \frac{1}{3}(3-x)$

  d $f(x) = \dfrac{1}{2x-5}$ e $f(x) = \dfrac{2}{3-4x}$ f $f(x) = 1 - \dfrac{1}{2x^3}$

9 $f:A \to A$ and $g:A \to A$, where $A = \{$positive real numbers$\}$, are functions defined by $f(x) = 2x$ and $g(x) = \dfrac{2}{3x}$. Find:

  a $f^{-1}(x)$ b $g^{-1}(x)$ c $(f^{-1} \circ g^{-1})(x)$ d $(g^{-1} \circ f^{-1})(x)$

10 Functions $u:R \to R$ and $v:R \to R$ are such that $u(x) = \frac{1}{2}x+1$ and $u(v(x)) = -x$. Find $v(x)$.

Cumulative Revision Exercises

Exercise 1

1   The breaking weight, W tonnes, of a rope of diameter d millimetres is proportional to the square of its diameter. Given that $W = 6$ when $d = 40$, find W when $d = 24$.

2   Sketch the graph of the function given by the formula $f(x) = (x-1)(x-2)$, $x \in R$. Hence find the interval of x over which the values of f are negative.

3   The sum of two real numbers, one of which is x, is 100. Prove that the product P of the numbers is given by $P = 2500 - (x-50)^2$. Write down the greatest value of P and the corresponding value of x.

4   Which of the following are true and which are false?

   a $x^{1/2} \times x^2 = x$ *b* $(x^{-2})^{1/2} = x^{-5/2}$ *c* $\dfrac{1}{4x} = 4x^{-1}$

5   Figure 2 shows the regions into which the axes and the lines $y = 1$ and $y = 4x$ divide the plane. Which regions show the following sets?

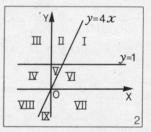

   a $P = \{(x, y) : y \geqslant 1\}$
   b $S = \{(x, y) : y \leqslant 4x\}$
   c $P \cap S$ *d* $(P \cup S)'$

6 *a*   Simplify: *(1)* $\sqrt{27} + \sqrt{12} - \sqrt{75}$ *(2)* $2\sqrt{2}(\sqrt{2} + \sqrt{8})$

   b   Taking $\sqrt{2} = 1\cdot414$, calculate $\dfrac{1}{2\sqrt{2}}$ to 2 decimal places.

7   Given $2x - y < 20 - 3x$, show that $y > 5(x-4)$. Hence find x for which $y > 20$.

8   Factorise: *a* $(p+q)^2 - r^2$ *b* $2x^2 - x - 10$ *c* $a^2b - 2ab^2 - 3b^3$

9   Solve, $x \in R$:   *a* $2x^2 - x - 6 = 0$ *b* $2x^2 - x - 5 = 0$.

Exercise 2

1   Simplify:

  a $(y^{1/2})^2 \times (a^{-2})^0$ b $(x^{1/2} + x^{-1/2})^2$ c $(3x)^{-1} - \dfrac{1}{3x}$

2   Find the solution sets of the following, $x \in R$:

  a $(x+5)(x-5) < 0$ b $(x-4)(2-x) \geqslant 0$ c $6+x-x^2 < 0$

3   ABCD is a square of side 10 cm. E is a point on AB such that
  BE $= x$ cm, and F is on BC such that CF $= 2x$ cm. D, E, F are joined.

  a   Write down expressions for the lengths of AE and BF.
  b   If the area of $\triangle$DEF is A cm², prove that $A = 25 + (x-5)^2$.
  c   Find the minimum value of A and the corresponding value of x.

4   Find the roots of the equation $\dfrac{x}{x+1} - \dfrac{x+1}{x} = 1$ to one decimal place.

5   $p = 3 + 2\sqrt{2}$ and $q = 3 - 2\sqrt{2}$. Find in their simplest forms:

  a $p+q$ b $p-q$ c p^2-q^2 d pq

6 a   Which regions in Figure 3 show the following?
      (1) $P = \{(x,y) : y \geqslant x+3\}$
      (2) $P \cap Q$, where $Q = \{(x,y) : y \geqslant -x+3\}$

  b   Give a set definition for the region containing
      VIII and IX only.

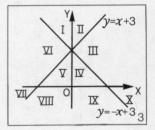

7   The stopping distances, d metres, of a car travelling at v m/s, are
    shown in the table.

v	5	15	30	60
d	5	25	80	280

  a   Assuming that d and v are connected by the formula $d = av^2 + bv$,
      where a and b are constants, find a and b, and state the formula.
  b   Use the formula to calculate d when $v = 45$, and v when $d = 40$.

8   Matrix P represents the geometrical transformation of a rotation of
    $90°$ anticlockwise about O, and matrix Q reflection in the line $y = x$.
  a   Find P, Q, PQ and QP.
  b   What transformations are represented by PQ and QP?
  c   Find matrix R such that $PQR = QP$, and describe the transformation
      given by R.

Exercise 3

1 a Find the set of points in which the line $y = x + 1$ intersects the circle $x^2 + y^2 = 25$.

   b Find c for which the line $y = x + c$ is a tangent to the circle.

2 The nth term of a sequence is given by $u_n = 6n - 30$.

   a What is the hundredth term?

   b Are 153 and 384 members of the sequence?

   c How many terms of the sequence are negative?

3 $M_1 = \begin{pmatrix} 1 & 1 \\ 0 & -1 \end{pmatrix}$, $M_2 = \begin{pmatrix} 1 & -1 \\ 1 & 0 \end{pmatrix}$, $M_3 = \begin{pmatrix} 0 & 1 \\ -1 & 1 \end{pmatrix}$, $M_4 = \begin{pmatrix} 1 & 0 \\ 0 & 1 \end{pmatrix}$.

Which of the following are false?

   a $M_1 M_2 = \begin{pmatrix} 2 & -1 \\ 1 & 0 \end{pmatrix}$ *b* $M_2 M_4 = \begin{pmatrix} 1 & -1 \\ 1 & 0 \end{pmatrix}$ *c* $M_2 M_3 = \begin{pmatrix} 1 & 0 \\ 0 & -1 \end{pmatrix}$

4 Functions $f : R \to R$ and $g : R \to R$ are defined by $f(x) = 1 - x^2$ and $g(x) = 2x - 1$.

   a Calculate $(f \circ g)(3)$ and $(g \circ f)(3)$.

   b Find formulae for $(f \circ g)(x)$ and $(g \circ f)(x)$, and check your answers for *a*.

   c Find a, given that $(f \circ g)(a) = -8$.

5 a Factorise the polynomials:

    (*1*) $x^3 - 3x^2 - 6x + 8$ (*2*) $x^3 - 13x - 12$

   b Solve the following equations, $x \in R$:

    (*1*) $2x^3 + 7x^2 + 2x - 3 = 0$ (*2*) $x^4 + 7x^3 + 11x^2 - 7x - 12 = 0$

6 Which of the following are true, and which are false?

   a $\log 3 + \log 3^2 = \log 3^3$ *b* $\log(5 - 3) = \log 5 - \log 3$

   c $\dfrac{\log 5^2}{\log 5} = \log 5$ *d* $\dfrac{\log 5^2}{\log 5} = 2$

7 Make n the subject of each of the following formulae:

   a $nE = I(r + Rn)$ *b* $\dfrac{1}{k} = \dfrac{1}{m} + \dfrac{1}{n}$ *c* $r = \dfrac{\sqrt{(1 - n^2)}}{n}$

Exercise 4

1 Factorise $x^3 - x^2 - 17x - 15$. Write down the coordinates of the points where the curve $y = x^3 - x^2 - 17x - 15$ cuts the x-axis, and find the gradient of the curve at each of these points.

2 *a* Find the sum of n terms and the sum to infinity of the series $\frac{1}{4} + \frac{1}{8} + \frac{1}{16} + \frac{1}{32} + \ldots$

 b Find the smallest value of n for which the sum of n terms and the sum to infinity differ by less than 0·001.

3 *a* For the function $f : x \rightarrow \dfrac{6}{x^2 - 1}$, $x \neq 1$ or -1, find:

  (1) the values of f at $x = -3, 0$ and $\frac{1}{2}$
  (2) the elements of the domain which have image 2

 b A function $g : R \rightarrow R$, defined by the formula $g(x) = ax + b$, is such that $g(6) = 4$ and $g(-2) = -8$. Find a, b and $g(4)$.

4 $P = \begin{pmatrix} a & 2 \\ -1 & b \end{pmatrix}$, $Q = \begin{pmatrix} -1 & -2 \\ 0 & 3 \end{pmatrix}$ and $PQ = \begin{pmatrix} -1 & 4 \\ 1 & -1 \end{pmatrix}$.

 a Find a and b, and hence obtain the matrix product QP.
 b Find the matrix M such that $MPQ = QP$.

5 *a* Simplify $\log_2 3 + \log_2 4 + \log_2 5 - \log_2 30$.
 b Calculate $\log_e 8\cdot 25$, where $e = 2\cdot72$.
 c If $\log_{10} p = k - \log_{10} q$, which one of the following is true?

  (1) $p = q10^k$ *(2)* $p = k/q$ *(3)* $p = k - 10^q$ *(4)* $pq = 10^k$

6 State the converse of each of the following. Then say whether each implication is true or false, and whether each converse is true or false.

 a If a quadrilateral is a kite, it has exactly one axis of symmetry.
 b If $810 < a < 990$, then $-1 \leqslant \cos a^\circ < 0$. (Use cosine graph.)
 c If $x^2 + 2x + 3 = 0$, then $x \notin R$.
 d If $\log_{10} p < 0$, then $0 < p < 1$.

Exercise 5

1 Find the solution sets of the following systems of equations $(x, y, z \in R)$:

a $\qquad 5x + 2y = 1$
$\qquad 4x^2 - 5xy - y^2 = 10$

b $\qquad 2x + y + 2z = 9$
$\qquad 5x - 2y - 3z = 3$
$\qquad 3x + 3y + 4z = 15$

2 Mappings $f: R \to R$ and $g: R \to R$ are defined by $f(x) = 2x$ and $g(x) = 3 - 5x$.

a Find formulae for: (1) $(g \circ f)(x)$ (2) $(g \circ f)^{-1}(x)$
(3) $f^{-1}(x)$ (4) $g^{-1}(x)$ (5) $(f^{-1} \circ g^{-1})(x)$

b Which two mappings are equal?

3 $A = \begin{pmatrix} 2 & 3 \\ -3 & 2 \end{pmatrix}$ and $B = \begin{pmatrix} 1 & 4 \\ -1 & 4 \end{pmatrix}$. Is it true that:

a $A + B = B + A$ b $AB = BA$ c $(A+B)(A-B) = A^2 - B^2$?

4 The function f is defined by the formula $f(x) = x^3 + x^2 - 5x + 3$.

a Express $f(x)$ as a product of linear factors, and state the zeros of f.
b Find the stationary values of f, and investigate the nature of each.
c Sketch the graph of f.

5 a Show that $\dfrac{1 + 2 + 3 + \ldots + (n-1)}{n + n + n + \ldots + n} = \dfrac{1}{2}$, where the numerator and denominator each contain $n-1$ terms.

b A hall has parallel rows of seats, each row having two seats fewer than the one behind it. If the longest row has 128 seats, and there are 2180 seats altogether, how many rows are there?

6 If α and β are the roots of $2x^2 + 4x + 1 = 0$, find the values of:

a $\dfrac{1}{\alpha} + \dfrac{1}{\beta}$ b $\alpha^2 + \beta^2$ c $\dfrac{1}{\alpha^2} + \dfrac{1}{\beta^2}$ d $\left(\alpha - \dfrac{1}{\alpha}\right)\left(\beta - \dfrac{1}{\beta}\right)$

7 The curve $y = kx^2 + 24x + 9k$ touches the x-axis at its maximum turning point. Find k, and sketch the curve.

8 Find the solution sets of the following inequations, $x \in R$:

a $x^2 - 5x - 6 < 0$ b $2 + x - x^2 \geqslant 0$ c $16 - x^2 < 0$

Exercise 6

1   Use a matrix method to solve these systems of equations $(x, y \in R)$:

a $\left.\begin{array}{l} 5x+3y = 25 \\ 4x+3y = 23 \end{array}\right\}$ b $\left.\begin{array}{l} 5x-3y = 29 \\ 7x+2y = 22 \end{array}\right\}$ c $\left.\begin{array}{l} px+qy = r \\ qx+py = s \end{array}\right\}, p^2 \neq q^2$

2 a   Show that 3 is a root of $2x^3 - 11x^2 + 12x + 9 = 0$, and find the other roots.

   b   Show that $x^3 + x - 1 = 0$ has a root between 0 and 1, and find it to one decimal place.

3 a   Find the inverse of each of the following functions from R to R:

   (1) $f: x \rightarrow 8 - x$ (2) $g: x \rightarrow 8x + 1$ (3) $h: x \rightarrow \frac{1}{2}(3x - 4)$

   b   Functions $F: R \rightarrow R$ and $G: R(\text{excluding } 0) \rightarrow R(\text{excluding } 0)$ are defined by $F(x) = x + 1$ and $G(x) = 1/x$. Which of the following are true and which are false?

   (1) $G(\frac{1}{2}) = F(1)$ (2) $F(F(x)) = x + 2$
   (3) $G(F(-1))$ is not defined   (4) $F(G(x)) = 1/(x + 1)$
   (5) $F(F^{-1}(x)) = x$ (6) $G^{-1}(G(x)) = x$

4   The third term of a geometric sequence is 32 and the sixth term is 2048. Find the first term and the common ratio.
     Show that the nth term of the sequence can be put in the form 2^{2n-1}, and that the *product* of the first n terms is 2^{n^2}.

5   The circle $x^2 + y^2 + 2gx + 2fy + c = 0$ passes through the points $(4, 4), (0, -4)$ and $(4, 0)$. Find g, f and c, and hence state the equation of the circle.

6 a   Find k if the roots of $(k+2)x^2 - 2kx + (k-1) = 0$ are:

   (1) equal   (2) not real.

   b   Which of the following gives the solutions of the equation $\log_2(p^2 - 14p) = 5$?

   (1) $7 \pm \sqrt{74}$ (2) $-2, 16$ (3) $7 \pm \sqrt{59}$ (4) $2, -16$

7   In the formula $I = \dfrac{E}{R}(1 - e^{-Rt/L})$, find $e^{-Rt/L}$ in terms of I, E and R.

     Hence obtain a formula to give t, and calculate t given that $E = 250$, $e = 2 \cdot 72$, $L = 2 \cdot 5$, $R = 50$ and $I = 1$.

Geometry

The Equations of a Circle

1

1 The equation of the circle, centre the origin, radius r

Exercise 1

1   Show on the Cartesian plane the following sets.

a $\{(x, y) : x^2 + y^2 = 25\}$ b $\{(x, y) : x^2 + y^2 > 25\}$

c $\{(x, y) : x^2 + y^2 < 25\}$

2   Describe algebraically (by means of an equation or an inequation) each of the following statements.

a   The point $P(x, y)$ is 3 units from the origin.
b $P(x, y)$ is more than 3 units from the origin.
c $P(x, y)$ is less than 3 units from the origin.

3   Describe the region of the Cartesian plane for which:

a $x^2 + y^2 = a^2$ b $x^2 + y^2 > a^2$ c $x^2 + y^2 < a^2$,

where a is a positive constant.

4   Describe algebraically each of the following statements.

a $P(x, y)$ is r units from the origin.
b $P(x, y)$ is more than r units from the origin.
c $P(x, y)$ is less than r units from the origin.

From questions *1* to *4* we see that the solution set of the open sentence $x^2 + y^2 = r^2$ consists of the coordinates of those points, and only those, which are r units from the origin, i.e. the circumference of the circle with centre the origin and radius r.

The equation $x^2+y^2=r^2$ of the circle with centre the origin and radius r

$P(x, y)$ is a point on the circle C with centre O and radius r. Then

$$C = \{P(x, y):OP = r\}$$
$$= \{P(x, y):OP^2 = r^2\}$$
$$= \{(x, y):x^2 + y^2 = r^2\}$$

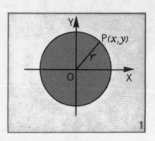

The equation $x^2 + y^2 = r^2$ is said to be the equation of the circle with centre O and radius r.

Note:

(i) If $P(a, b)$ lies on this circle then $a^2 + b^2 = r^2$. Conversely, if $a^2 + b^2 = r^2$ then $P(a, b)$ lies on the circumference.
Hence $P(a, b)$ lies on the circle $\Leftrightarrow a^2 + b^2 = r^2$.
(ii) $P(a, b)$ lies outside the circle $\Leftrightarrow a^2 + b^2 > r^2$.
(iii) $P(a, b)$ lies inside the circle $\Leftrightarrow a^2 + b^2 < r^2$.

Example 1. Find the equation of the circle, centre O, passing through the point $A(7, 1)$.

Since $A(7, 1)$ is on the circumference, $OA^2 = r^2$.
Hence $r^2 = 7^2 + 1^2 = 50$, and the equation of the circle is $x^2 + y^2 = 50$.

Example 2. A is the point $(0, 1)$ and B is $(0, 9)$. $P(x, y)$ belongs to a locus given by $\{P(x, y):PB = 3PA\}$. Describe this locus geometrically.

$$\{P(x, y):PB = 3\,PA\}$$
$$= \{P(x, y):PB^2 = 9\,PA^2\}$$
$$= \{(x, y):x^2 + (y-9)^2 = 9[x^2 + (y-1)^2]\}$$
$$= \{(x, y):x^2 + y^2 - 18y + 81 = 9x^2 + 9y^2 - 18y + 9\}$$
$$= \{(x, y):72 = 8x^2 + 8y^2\}$$
$$= \{(x, y):x^2 + y^2 = 9\}$$

The locus is the circle, centre O, radius 3 units (as shown in Figure 1, with $r = 3$).

Exercise 2

1   Write down the equations of the circles with centre O and radii:

a 3   *b* 5   *c* 8   *d* 9   *e* 1·2   *f* *a* units

2   Find the equations of the circles with centre the origin and passing through the points:

a $(2, 3)$ *b* $(-1, 2)$ *c* $(4, 0)$ *d* $(-6, -8)$

3   State the centre and radius of each of these circles:

a $x^2 + y^2 = 36$ *b* $x^2 + y^2 = 12$ *c* $4x^2 + 4y^2 = 9$

4   Find the equation of the circle with the same centre as the circle $x^2 + y^2 = 25$, but with radius twice as long.

5   The sides of a square have equations $x = 4$, $x = -4$, $y = 4$, $y = -4$. Find the equation of the circle:

a touching all the sides of the square
b passing through all the vertices of the square.

6   Find k if the given point lies on the given circle in each case.

a $(k, -2)$, $x^2 + y^2 = 13$ *b* $(k, 7)$, $x^2 + y^2 = 58$

c $(-5, k)$, $x^2 + y^2 = 41$ *d* (k, k), $x^2 + y^2 = 50$

7   Without drawing diagrams, state the position of each of the following points with respect to the regions P, Q, R of the x, y plane:

$P = \{(x, y) : x^2 + y^2 < 25\}$, $Q = \{(x, y) : x^2 + y^2 = 25\}$,
$R = \{(x, y) : x^2 + y^2 > 25\}$.

a $(3, 4)$ *b* $(2, 5)$ *c* $(-5, 0)$ *d* $(-2, -3)$

8   Which of these are true, and which are false, for the circle $x^2 + y^2 = 16$?

a The point $(0, -4)$ lies on the circle.
b The radius of the circle is 16 units in length.
c The point $(4\cos\theta, 4\sin\theta)$ lies on the circle, for all θ.
d The circle is concentric with the circle $2x^2 + 2y^2 = 1$.
e The circle can be inscribed in a square of area 16 square units.

9   Calculate the area enclosed between the circles $x^2 + y^2 = 8$ and $x^2 + y^2 = 18$.

10 A is $(1, 0)$ and B is $(9, 0)$. Find $\{P(x, y) : PB = 3\,PA\}$, and describe this locus.

11 Repeat question *10* for the following cases.

  a $A(0, 1), B(0, 4); \{P : PB = 2\,PA\}$
  b $A(0, 2), B(0, 8); \{P : PB = 2\,PA\}$
  c $A(0, -1), B(0, -25); \{P : PB = 5\,PA\}$
  d $A(1, 1), B(9, 9); \{P : PB = 3\,PA\}$

---

2 The equation of the circle, centre (a, b), radius r

Exercise 3

Reminder. If d is the distance between the points (x_1, y_1) and (x_2, y_2) then $d^2 = (x_2 - x_1)^2 + (y_2 - y_1)^2$.

1 Where on the x, y plane will you find all the points whose coordinates belong to the solution set of:

  a $(x - 5)^2 + (y - 6)^2 = 9$
  b $(x - 5)^2 + (y - 6)^2 > 9$
  c $(x - 5)^2 + (y - 6)^2 < 9$?

2 Describe algebraically each of the following statements.

  a The point $P(x, y)$ is 5 units from the point $A(8, 3)$.
  b $P(x, y)$ is more than 5 units from $A(8, 3)$.
  c $P(x, y)$ is less than 5 units from $A(8, 3)$.

3 For a given point $A(a, b)$, and a positive constant c, describe the region of the Cartesian plane for which:

  a $(x - a)^2 + (y - b)^2 = c^2$
  b $(x - a)^2 + (y - b)^2 > c^2$
  c $(x - a)^2 + (y - b)^2 < c^2$.

4 Give an algebraic statement for each of the following.

  a $P(x, y)$ is r units from $A(a, b)$.
  b $P(x, y)$ is more than r units from $A(a, b)$.
  c $P(x, y)$ lies within the circle, centre $A(a, b)$, radius r units.

From questions *1* to *4* we see that the solution set of the open sentence $(x-a)^2+(y-b)^2 = r^2$ consists of the coordinates of those points, and only those, which are r units distant from the point A(a, b), i.e. the circumference of the circle, centre A, radius r.

The equation $(x-a)^2+(y-b)^2 = r^2$ of the circle with centre (a,b) and radius r

P(x, y) is a point on the circle C with centre A(a, b) and radius r. Then

$C = \{P(x, y) : AP = r\}$
$\quad = \{P(x, y) : AP^2 = r^2\}$
$\quad = \{(x, y) : (x-a)^2+(y-b)^2 = r^2\}$

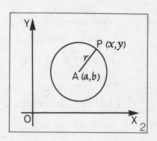

The equation $(x-a)^2+(y-b)^2 = r^2$ is said to be the equation of the circle with centre (a, b) and radius r.

Note:

(i) P(h, k) lies on the circle $\quad\Leftrightarrow (h-a)^2+(k-b)^2 = r^2$.
(ii) P(h, k) lies outside the circle $\Leftrightarrow (h-a)^2+(k-b)^2 > r^2$.
(iii) P(h, k) lies inside the circle $\quad\Leftrightarrow (h-a)^2+(k-b)^2 < r^2$.

Exercise 4

In this Exercise, give the equations of the circles in the form
$$(x-a)^2+(y-b)^2 = r^2.$$

1　Write down the equations of the circles with the following centres and radii:

a　$(1, 1), 3$　　　*b*　$(0, 3), 4$　　　*c*　$(5, 0), 2$　　　*d*　$(-5, 2), 7$

e　$(-3, -4), 6$　*f*　$(2, -3), 1$　　*g*　$(-1, 0), 6$　*h*　$(-6, -6), 3$

2　Find the equations of the circles with given centres, and passing through the given points.

a　centre $(6, 8)$, through O　　　*b*　centre $(-8, 15)$, through O

c　centre $(1, 1)$, through $(3, 3)$　　*d*　centre $(-2, 0)$, through $(3, 4)$

3   State the centre and radius of each of these circles:

a $(x-1)^2+(y-3)^2 = 25$ b $(x-2)^2+(y+4)^2 = 49$

c $(x+1)^2+(y+8)^2 = 9$ d $(x+5)^2+y^2 = 1$

4   Find the equation of the circle concentric with the circle $(x+2)^2+(y-4)^2 = 25$, but with radius twice as long.

5   Write down the centres and equations of the four circles which touch both the x- and y-axes, and have radius 3.

6   Without drawing diagrams, find which of these points are inside, which are outside, and which are on the circle $(x-1)^2+(y+2)^2 = 16$:

a $(1, 2)$ b $(3, 1)$ c $(-3, -3)$ d $(5, -2)$

7   Two circles have centre C(1, 2). One of them passes through A(2, 4) and the other passes through B(3, 6). Find the ratio of:

a   the radii   b   the areas, of the circles

8 a   State the equation of the circle, centre (1, 5), touching the x-axis.
  b   State the equation of the image of this circle under reflection in:

    (1) the x-axis (2) the y-axis (3) the origin (4) the line $y = x$.

9 a   State the equation of the circle, centre (2, -1), touching the y-axis.
  b   State the equation of the image of this circle in:

    (1) the x-axis (2) the y-axis (3) the origin (4) the line $x+y = 0$.

10   Find the centres and radii, and hence the equations, of the circles on the lines joining the following pairs of points as diameters.

a   O(0, 0) and A(4, 6)   b   B(-3, 5) and C(1, -1)

11   A is the point (1, 2), B is (5, 6), C(1, 6). Show that angle ACB is a right angle, and find the equation of the circle passing through A, B and C.

12   A(x_1, y_1) and B(x_2, y_2) are the ends of a diameter of a circle, and P(x, y) is a point on the circumference. Write down the gradients of PA and PB, and hence show that the equation of the circle on AB as diameter can be written in the form

$$(x-x_1)(x-x_2)+(y-y_1)(y-y_2) = 0.$$

3   The general equation of a circle

So far we can recognise that the equation $(x-1)^2+(y-2)^2 = 16$ is the equation of the circle with centre $(1, 2)$ and radius 4 units.

Consider what happens if we expand $(x-1)^2$ and $(y-2)^2$, and rearrange terms.

If P is a point (x, y) and C is the point $(1, 2)$, then the circle with centre $C(1, 2)$ and radius 4 is

$$\{P(x, y):CP = 4\}$$
$$= \{P(x, y):CP^2 = 16\}$$
$$= \{\ (x, y):(x-1)^2+(y-2)^2 = 16\}$$
$$= \{\ (x, y):x^2-2x+1+y^2-4y+4 = 16\}$$
$$= \{\ (x, y):x^2+y^2-2x-4y-11 = 0\}$$

We can now say that the equation $x^2+y^2-2x-4y-11 = 0$ is the equation of the circle with centre $(1, 2)$ and radius 4.

From the equality of sets this process can easily be reversed, as in the following Example.

Example. Show that $x^2+y^2+2x-4y-20 = 0$ is the equation of a circle, and give its centre and radius.

$$\{P(x, y):x^2+y^2+2x-4y-20 = 0\}$$
$$= \{P(x, y):x^2+2x+y^2-4y = 20\}$$
$$= \{P(x, y):x^2+2x+1+y^2-4y+4 = 1+4+20\}$$
$$= \{P(x, y):(x+1)^2+(y-2)^2 = 25\}$$
$$= \{P(x, y):CP^2 = 25\}, \text{ where C is the point } (-1, 2), \text{ which is the}$$
circle with centre $(-1, 2)$ and radius 5.

Exercise 5

Use the method of completing the square to find the centre and radius of each of the following circles. In each case, make a rough sketch of the circle and note anything of particular interest such as a circle passing through the origin or touching an axis.

1   $x^2+y^2-2x-6y+6 = 0$ 2 $x^2+y^2-4x-4y = 0$

3   $x^2+y^2-6x-2y+9 = 0$ 4 $x^2+y^2+2x+2y+1 = 0$

5   $x^2+y^2+6x = 0$ 6 $x^2+y^2-4y = 0$

7   $x^2+y^2-6x-8y = 0$ 8 $x^2+y^2+2ax = 0, a \neq 0$

The equation $x^2+y^2+2gx+2fy+c=0$, where g, f, c are constants, is the equation of a circle, centre $(-g, -f)$, radius $\sqrt{(g^2+f^2-c)}$

Proof. $\{P(x, y):x^2+y^2+2gx+2fy+c = 0\}$
$= \{P(x, y):x^2+2gx+y^2+2fy = -c\}$
$= \{P(x, y):x^2+2gx+g^2+y^2+2fy+f^2 = g^2+f^2-c\}$
$= \{P(x, y):(x+g)^2+(y+f)^2 = g^2+f^2-c\}$
$= \{P(x, y):(x-(-g))^2+(y-(-f))^2 = g^2+f^2-c\}$
$= \{P(x, y):CP^2 = g^2+f^2-c\}$, where C is the point $(-g, -f)$,

which is the circle with centre $(-g, -f)$ and radius $\sqrt{(g^2+f^2-c)}$.

Note. $(x-a)^2+(y-b)^2 = r^2$ (where $a, b, r \in R$) represents the same set of loci as $x^2+y^2+2gx+2fy+c = 0$ (where $g, f, c \in R$) if and only if $g^2+f^2-c \geqslant 0$.

Example. Find the centre and radius of the circle which has equation $x^2+y^2-4x+6y-12 = 0$.
Compare $x^2+y^2+2gx+2fy+c = 0$.
Here $g = -2, f = 3$ and $c = -12$.
The centre is $(-g, -f)$, i.e. $(2, -3)$, and the radius is $\sqrt{(g^2+f^2-c)}$, i.e. $\sqrt{(4+9+12)} = \sqrt{25} = 5$.
Note. An equation in the form $3x^2+3y^2-4x+8y-1 = 0$ must first be rewritten as $x^2+y^2-\frac{4}{3}x+\frac{8}{3}y-\frac{1}{3} = 0$.

Exercise 6

1   Find the centre and radius of each of these circles:

a $x^2+y^2-2x-6y-15 = 0$ b $x^2+y^2+4x+2y+1 = 0$

c $x^2+y^2-4x+8y-5 = 0$ d $x^2+y^2-4x-4y+7 = 0$

2   Find the centre and radius of each of the following circles.

a $4x^2+4y^2-16x+8y+11 = 0$ b $2x^2+2y^2-4x+3y = 0$

c $x^2+y^2-2x\cos\theta-2y\sin\theta+\sin^2\theta = 0$ (Answer in terms of θ.)

3 a Find the centre and radius of the circle $x^2+y^2+8x-2y = 8$.
  b Use the theorem of Pythagoras to calculate the length of the tangent to the circle from the point $(2, 4)$.

4   Calculate the length of the tangent from the point $(3, 2)$ to the circle $x^2+y^2+4x-2y = 13$.

5   Which of the following are equations of circles?

a $2x + 3y - 4 = 0$ *b* $x^2 + y^2 - 5 = 0$

c $x^2 + y^2 - x + y - 2 = 0$ *d* $2x^2 + 3y^2 + x - y = 0$

e $(x-1)^2 + (y-2)^2 = 6$ *f* $x^2 + y^2 + xy - x - y + 1 = 0$

6   Find h if the point $(h, 3)$ lies on the circle $x^2 + y^2 + 13x + 5y + 6 = 0$.

7   Find k if the point $(-5, k)$ lies on the circle $x^2 + y^2 + 2x - 5y - 21 = 0$.

8   Which of these points lie on the circle $x^2 + y^2 + 4x - 8y - 5 = 0$?

a $(1, 8)$ *b* $(8, 1)$ *c* $(-2, -1)$ *d* $(2, 1)$ *e* $(-2, 9)$

9   Find g if the point $(1, 2)$ lies on the circle $x^2 + y^2 + 2gx + 3y + 1 = 0$.

10   Find f if the point $(-1, 2)$ lies on the circle $x^2 + y^2 - 5x + 2fy - 6 = 0$.

11   Find g, f and c if the circle $x^2 + y^2 + 2gx + 2fy + c = 0$ passes through the origin and the points $(1, 3)$ and $(5, -5)$.

12a Find the equation of the circle through the points $O(0, 0)$, $A(-2, 4)$ and $B(-1, 7)$.

  b Find also the equation of the circle concentric with the first circle, and passing through the midpoint of OA.

13   Find the equation of the circle passing through the points $A(0, -1)$, $B(2, 3)$ and $C(1, 6)$, i.e. the circumcircle of $\triangle$ABC.

14   Find the equation of the circumcircle of the triangle with vertices $(2, 1), (1, 2), (1, 0)$.

15   A is the point $(-3, 4)$ and B is $(2, -1)$. Show that the equation of the locus of $P(x, y)$ given by $\{P : 2AP = 3PB\}$ is a circle, and find the centre and radius of the circle.

---

4   *The intersection of a line and a circle*

Exercise 7

1 $C = \{(x, y) : x^2 + y^2 = 4\},$ $L_1 = \{(x, y) : x = 1\},$
$L_2 = \{(x, y) : x = 2\},$ $L_3 = \{(x, y) : x = 3\}.$

Show clearly in a diagram: *a* $C \cap L_1$ *b* $C \cap L_2$ *c* $C \cap L_3$.
Which line is a tangent to the circle?

2 Find the solution sets of the following systems of equations by an algebraic method, and then interpret the results geometrically.

a $\left.\begin{aligned} x^2+y^2 &= 9 \\ y &= 1 \end{aligned}\right\}$ b $\left.\begin{aligned} x^2+y^2 &= 9 \\ y &= 2 \end{aligned}\right\}$ c $\left.\begin{aligned} x^2+y^2 &= 9 \\ y &= 3 \end{aligned}\right\}$ d $\left.\begin{aligned} x^2+y^2 &= 9 \\ y &= 4 \end{aligned}\right\}$

3 Find the points of intersection of these circles with the x- and y-axes.

a $x^2+y^2 = 4$ b $x^2+y^2 = 36$ c $x^2+y^2 = 64$

4 Find the points of intersection of these circles with the lines $y = x$ and $x+y = 0$.

a $x^2+y^2 = 8$ b $x^2+y^2 = 32$ c $x^2+y^2 = 98$

5 $L = \{(x, y) : y = x+4\}$ and $C = \{(x, y) : x^2+y^2 = 4\}$. Show that $L \cap C = \phi$, and interpret the result geometrically.

6 Find the solution set of each of the following systems of equations, and illustrate by sketches.

a $\left.\begin{aligned} x^2+y^2 &= 20 \\ y &= 2x \end{aligned}\right\}$ b $\left.\begin{aligned} x^2+y^2 &= 16 \\ y &= x-4 \end{aligned}\right\}$ c $\left.\begin{aligned} x^2+y^2 &= 8 \\ x+y &= 4 \end{aligned}\right\}$

7 Find the points of intersection of these lines and circles:

a $\left.\begin{aligned} x^2+y^2+2x+2y+1 &= 0 \\ y-x &= 1 \end{aligned}\right\}$ b $\left.\begin{aligned} x^2+y^2+4x+2y &= 0 \\ 2x-y+8 &= 0 \end{aligned}\right\}$

8 Show that these lines are tangents to the given circles:

a $\left.\begin{aligned} x^2+y^2 &= 2 \\ y &= x+2 \end{aligned}\right\}$ b $\left.\begin{aligned} x^2+y^2+10x+4y+19 &= 0 \\ x+3y+1 &= 0 \end{aligned}\right\}$

9 Find the equation of the circle, centre A(2, 1), which passes through the origin. A chord BC of the circle has midpoint M($\frac{1}{2}$, $2\frac{1}{2}$). Find the coordinates of B and C.

10 Show that the equation of the circle with centre in the first quadrant, radius k and touching the y-axis at (0, 3), is $(x-k)^2+(y-3)^2 = k^2$.

If the circle passes through (2, 7) find k. Find also the equation of the circle and the length of the chord cut off by the x-axis.

5 *Tangents to a circle*

(i) The tangent at a given point on a circle

Example. Show that A(5, 1) lies on the circle $x^2 + y^2 - 4x + 6y - 12 = 0$. If C is the centre of the circle find the gradient of CA and hence the gradient of the tangent at A. Deduce the equation of the tangent at A.

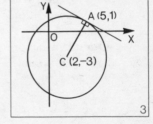

At $x = 5$, $y = 1$,
$$x^2 + y^2 - 4x + 6y - 12$$
$$= 25 + 1 - 20 + 6 - 12$$
$$= 0$$
Hence A(5, 1) lies on the circle.

C is $(2, -3)$, so $m_{CA} = \dfrac{1+3}{5-2} = \dfrac{4}{3}$.

It follows that the gradient of the tangent is $-\frac{3}{4}$, and that the equation of the tangent is

$$y - 1 = -\tfrac{3}{4}(x - 5)$$
$$\Leftrightarrow \quad 4y - 4 = -3x + 15$$
$$\Leftrightarrow 3x + 4y = 19$$

Exercise 8

1   Verify that the point $(-3, 4)$ lies on the circle $x^2 + y^2 = 25$, and find the equation of the tangent to the circle at this point.

2   Show that $(1, -3)$ lies on the circle $x^2 + y^2 = 10$, and find the equation of the tangent to the circle at this point.

3   Find the equation of the tangent to the circle $x^2 + y^2 = 13$ at the point $(-2, 3)$.

4   Find the equation of the tangent to the circle $x^2 + y^2 = 8$ at each of these points:

  a (2, 2)   *b* (2, -2)   *c* (-2, -2)   *d* (-2, 2)

5   Find the equation of the tangent to the circle $(x - 1)^2 + (y - 5)^2 = 9$ at the point (1, 2).

6   Find the equation of the tangent at the point $(4, -1)$ on the circle $x^2 + y^2 + 6x - 4y - 45 = 0$.

7   Find the equation of the tangent at the point $(-1, 2)$ on the circle $3x^2 + 3y^2 - 6x - 9y - 3 = 0$.

8   Find the equations of the tangents to the circle $x^2 + y^2 = 25$ at the points A and B where the line $x = 3$ cuts the circle. Find the coordinates of the point of intersection of the tangents at A and B.

9 *a*   Find the coordinates of the three points where the circle, centre $(4, 3)$, radius 5, cuts the axes.

   b   Find the equations of the tangents to the circle at these three points, and show that two of them are parallel.

   c   Write down the coordinates of the fourth point on the circle at which the tangent completes a parallelogram with the three already found.

(ii) Conditions for tangency

We saw in Section 4 and also in Algebra, Chapter 1, that we can interpret the intersection of a line and a circle both geometrically and algebraically.

Line meets circle:	Quadratic equation has:	Discriminant:
(i) in two distinct points	two distinct real roots	$b^2 - 4ac > 0$
(ii) in two coincident points	two equal roots	$b^2 - 4ac = 0$
(iii) in no points	no real roots	$b^2 - 4ac < 0$

Example. Find the equations of the tangents to the circle $x^2 + y^2 = 9$ from the point $(0, -5)$ outside the circle.

Using the equation $y - b = m(x - a)$, for comparison, we see that the line through $(0, -5)$ with gradient m

is $y + 5 = m(x - 0)$

$\Leftrightarrow \qquad y = mx - 5$.   .   .   (1)

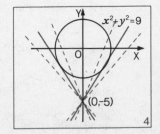

This line cuts the circle $x^2 + y^2 = 9$ at points whose x-coordinates are given by

$$x^2 + (mx - 5)^2 = 9$$
$$x^2 + m^2x^2 - 10mx + 25 = 9$$
$$(1 + m^2)x^2 - 10mx + 16 = 0 \qquad (2)$$

Line (1) is a tangent if and only if it meets the circle in two coincident points, i.e. if and only if the quadratic equation (2) has equal roots,

$$\Leftrightarrow \qquad \text{`}b^2 - 4ac = 0\text{'}$$
$$\Leftrightarrow (-10m)^2 - 4.16(1 + m^2) = 0$$
$$\Leftrightarrow \qquad 100m^2 - 64 - 64m^2 = 0$$
$$\Leftrightarrow \qquad\qquad 36m^2 = 64$$
$$\Leftrightarrow \qquad\qquad m = \pm\tfrac{4}{3}$$

From (1), the tangents are $y = \tfrac{4}{3}x - 5$ and $y = -\tfrac{4}{3}x - 5$, i.e. $4x - 3y - 15 = 0$ and $4x + 3y + 15 = 0$.

Exercise 9

1 Show that the line $y = 0$ is a tangent to the circle $x^2 + y^2 - 4x - 2y + 4 = 0$, and find the point of contact.

2 Show that the line $y = 3x + 10$ is a tangent to the circle $x^2 + y^2 - 8x - 4y - 20 = 0$, and find the point of contact.

3 Prove that the y-axis is a tangent to the circle $x^2 + y^2 - 2ax \cos \theta - 2ay \sin \theta + a^2 \sin^2 \theta = 0$.

4 Find the value of r^2 if the line $y = 2x - 5$ is a tangent to the circle $x^2 + y^2 = r^2$.

5 *a* Find the gradient of the radius through the point A(8, −6) on the circle $x^2 + y^2 + 20y + 20 = 0$, and hence find the equation of the tangent at A to the circle.

  b Prove that this tangent is also a tangent to the circle $x^2 + y^2 = 20$.

6 *a* Find the equation of the tangent at A(−3, −1) to the circle $x^2 + y^2 = 10$, and show that this tangent is also a tangent to the circle $x^2 + y^2 - 20y + 60 = 0$, giving its point of contact.

  b By considering the y-axis as an axis of bilateral symmetry, write down the equation of the other direct common tangent to the circles and the coordinates of the points of contact.

7 *a* Show that the line $y = kx$ meets the circle $x^2 + y^2 - 4x - 2y + 4 = 0$ at points whose x-coordinates are given by the equation $(1 + k^2)x^2 - 2(2 + k)x + 4 = 0$.

b   Hence find the values of k for which:

(1) the line is a tangent to the circle
(2) the line meets the circle in two distinct points
(3) the line does not cut the circle.

8   Repeat question 7 for the line $y = kx$ and the circle $2x^2 + 2y^2 - 8x + 4y + 5 = 0$.

9   Write down the equation of the line through the point $(0, 2)$, with gradient m. Hence show that the equations of the tangents from the point $(0, 2)$ to the circle $x^2 + y^2 = 1$ are $y = \sqrt{3}x + 2$ and $y = -\sqrt{3}x + 2$.

10   Find the equations of the tangents to the circle $x^2 + y^2 = 9$ from the point $(0, 5)$ outside the circle.

11   A is $(-8, 0)$, B$(-2, 0)$. A locus is described by $\{P(x, y) : PA = 2PB\}$. Find the equation of the locus, and describe it geometrically. Find the equations of the tangents to the locus from the point $(0, -5)$.

12   The straight line $y = -2x + c$ is a tangent to the circle $x^2 + y^2 - 2x - 4y - 15 = 0$. Calculate the possible values of c.

13a A circle, centre A(4, 3), passes through the origin; find the equation of the circle.

b   A chord PQ of the circle has midpoint M(2, 2); find P and Q.

c   If R is $(-6, -2)$, show that RP and RQ are tangents to the circle.

14   Show that the tangent to the circle $x^2 + y^2 = 10$ at $(-1, -3)$ is also a tangent to the circle $x^2 + y^2 + 4x - 8y = 20$, and find the length of this common tangent.

Summary

1 *The circle, centre* O, *radius r*, has equation

$$x^2 + y^2 = r^2.$$

2 *The circle, centre* (a, b), *radius r*, has equation

$$(x - a)^2 + (y - b)^2 = r^2.$$

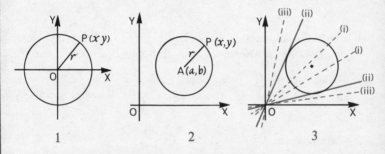

1 2 3

3 *The general equation of a circle* is

$$x^2 + y^2 + 2gx + 2fy + c = 0 \qquad (g^2 + f^2 - c \geqslant 0).$$

The centre is $(-g, -f)$, and the radius has length $\sqrt{(g^2 + f^2 - c)}$.

4

Line meets circle:	*Quadratic equation has*:	*Discriminant*:
(i) in two distinct points	two distinct real roots	$b^2 - 4ac > 0$
(ii) in two coincident points	two equal roots	$b^2 - 4ac = 0$
(iii) in no points	no real roots	$b^2 - 4ac < 0$

(See Figure 3)

Composition of Transformations 2

1   *Matrices associated with transformations*

(i) Revision

We have already seen in Books 5 and 7 that 2×2 matrices can be associated with transformations of all points in the Cartesian plane.

Example 1. Find the product $\begin{pmatrix} 0 & -1 \\ 1 & 0 \end{pmatrix}\begin{pmatrix} a \\ b \end{pmatrix}$, and state the transformation associated with the matrix $\begin{pmatrix} 0 & -1 \\ 1 & 0 \end{pmatrix}$.

$$\begin{pmatrix} 0 & -1 \\ 1 & 0 \end{pmatrix}\begin{pmatrix} a \\ b \end{pmatrix} = \begin{pmatrix} -b \\ a \end{pmatrix}.$$

Hence $(a, b) \rightarrow (-b, a)$, which is the image of (a, b) under a positive (anticlockwise) rotation of $\frac{1}{2}\pi$ radians about the origin, as shown in Figure 1. Since this is true for every point in the plane, the transformation is a rotation of $\frac{1}{2}\pi$ radians about O.

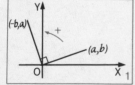
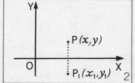
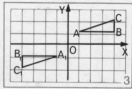

Example 2. Find the matrix associated with the transformation of reflection in the *x*-axis.

Under reflection in the x-axis, the point $P(x, y)$ maps to the point $P(x_1, y_1)$, where $x_1 = x$ and $y_1 = -y$ (see Figure 2).

Hence $\qquad\qquad x_1 = x = 1x + 0y$

and $\qquad\qquad y_1 = -y = 0x - 1y.$

So we can write $\begin{pmatrix} x_1 \\ y_1 \end{pmatrix} = \begin{pmatrix} 1 & 0 \\ 0 & -1 \end{pmatrix} \begin{pmatrix} x \\ y \end{pmatrix}$, which shows that the required matrix is $\begin{pmatrix} 1 & 0 \\ 0 & -1 \end{pmatrix}$.

Example 3. Find the image of $\triangle ABC$, with vertices $A(1, 1)$, $B(4, 1)$ and $C(4, 2)$ under the transformation with matrix $\begin{pmatrix} -1 & 0 \\ 0 & -1 \end{pmatrix}$.

The image of every point under the transformation can be found by writing the coordinates of the point as a column matrix, and premultiplying the resulting matrix by $\begin{pmatrix} -1 & 0 \\ 0 & -1 \end{pmatrix}$.

$$\begin{array}{ccc} A & B & C \end{array} \qquad\qquad \begin{array}{ccc} A_1 & B_1 & C_1 \end{array}$$
$$\begin{pmatrix} -1 & 0 \\ 0 & -1 \end{pmatrix} \begin{pmatrix} 1 & 4 & 4 \\ 1 & 1 & 2 \end{pmatrix} = \begin{pmatrix} -1 & -4 & -4 \\ -1 & -1 & -2 \end{pmatrix}$$

$\triangle A_1 B_1 C_1$, with vertices $A_1(-1, -1)$, $B_1(-4, -1)$ and $C_1(-4, -2)$, is the image of $\triangle ABC$ under the given transformation, which is a half turn about O (see Figure 3).

We have assumed that straight lines map to straight lines, e.g. $AB \to A_1 B_1$; this is investigated more fully in Section 5.

Exercise 1

1 a Simplify $\begin{pmatrix} 0 & 1 \\ 1 & 0 \end{pmatrix} \begin{pmatrix} a \\ b \end{pmatrix}$, and state the transformation associated with the matrix $\begin{pmatrix} 0 & 1 \\ 1 & 0 \end{pmatrix}$.

b Repeat *a* for the matrix $\begin{pmatrix} 0 & -1 \\ -1 & 0 \end{pmatrix}$.

2 As in Worked Example 2 above, illustrate each of the following transformations on a diagram, and find the associated matrix:

a reflection in the *y*-axis
b rotation of $-\frac{1}{2}\pi$ radians (i.e. clockwise) about O.

3 Copy and complete the following table. Keep it for reference.

Transformation	Matrix	Transformation	Matrix
Identity	$\begin{pmatrix} 1 & 0 \\ 0 & 1 \end{pmatrix}$	Reflection in $y = x$	$\begin{pmatrix} 0 & 1 \\ 1 & 0 \end{pmatrix}$
Half turn about O		Reflection in $y = -x$	
Reflection in *x*-axis		Rotation about O of $\frac{1}{2}\pi$ radians	
Reflection in *y*-axis		Rotation about O of $-\frac{1}{2}\pi$ radians	

4 A(1, 3), B(3, −2) and C(6, 4) are the vertices of △ABC. Use the appropriate matrices to calculate the coordinates of the images of A, B and C under:

a reflection in the line $y = x$
b rotation of $\frac{1}{2}\pi$ radians about O.

5 The vertices of a rhombus are P(−1, −1), Q(2, 0), R(3, 3) and S(0, 2). Use matrices to calculate the coordinates of the images of these vertices under:

a reflection in the *y*-axis
b reflection in the line $y = -x$.

Illustrate *a* and *b* in separate diagrams.

6 Under a mapping, P(*x*, *y*) → P$_1$(x_1, y_1), where $x_1 = x$ and $y_1 = -y$.

a State the associated matrix, and describe the mapping geometrically.
b Find the image of the set $\{(1, 2), (5, -10), (a, b), (c, d)\}$.
c Show that the distances between the pair of points (a, b), (c, d) and their pair of images are equal.

7 A geometrical transformation is associated with the matrix $\begin{pmatrix} 1 & 2 \\ 0 & 1 \end{pmatrix}$.

a Find the images of the vertices of rectangle O(0, 0), A(2, 0), B(2, 1), C(0, 1).
b Show the rectangle and its image in a diagram.
c Is distance preserved in this mapping?
d Is the area of the rectangle equal to the area of its image?

8   Under a mapping, $P(x,y) \rightarrow P_1(x_1, y_1)$, where $x_1 = 3x - 4y$ and $y_1 = 4x + 3y$.

   a   What matrix is associated with this mapping?

   b   Find the images of the vertices of the unit square $O(0,0)$, $A(1,0)$, $B(1,1)$, $C(0,1)$.

   c   Illustrate by a diagram on squared paper, and interpret the mapping as the composition of two geometrical transformations.

9 *a*   Find the images A_1 and B_1 of the points $A(2, -1)$ and $B(-4, 5)$ under the transformation with matrix $\begin{pmatrix} 3 & 2 \\ 1 & 4 \end{pmatrix}$.

   b   Show that the image of the midpoint of AB is the midpoint of $A_1 B_1$.

10   $P(a,b) \rightarrow P_1(7,2)$ under the transformation with matrix $\begin{pmatrix} 4 & 1 \\ -1 & 1 \end{pmatrix}$. Find a and b.

11   Under a transformation with associated 2×2 matrix $\begin{pmatrix} a & b \\ c & d \end{pmatrix}$, $(2,1) \rightarrow (5,1)$ and $(0,1) \rightarrow (1,3)$. Find a, b, c and d.

(ii) Matrices associated with dilatation

Exercise 2

1 *a*   Write down the coordinates of the image of the point (a, b) under the dilatation $[O, 3]$.

   b   Calculate the product $\begin{pmatrix} 3 & 0 \\ 0 & 3 \end{pmatrix}\begin{pmatrix} a \\ b \end{pmatrix}$.

2 *a*   Write down the coordinates of the image of the point (a, b) under the dilatation $[O, -2]$.

   b   Find the 2×2 matrix which transforms $\begin{pmatrix} a \\ b \end{pmatrix}$ into $\begin{pmatrix} -2a \\ -2b \end{pmatrix}$.

3   Find the 2×2 matrix which transforms $\begin{pmatrix} a \\ b \end{pmatrix}$ into $\begin{pmatrix} ka \\ kb \end{pmatrix}$, $k \neq 0$.

4   Find the image set of $\{(1, 3), (-2, 4), (3, -5), (a, b)\}$ under the transformation associated with the matrix:

a $\begin{pmatrix} 2 & 0 \\ 0 & 2 \end{pmatrix}$ b $\begin{pmatrix} k & 0 \\ 0 & k \end{pmatrix}$, k non-zero.

The matrix associated with the dilatation $[O, k]$ is $\begin{pmatrix} k & 0 \\ 0 & k \end{pmatrix}$. Note that the centre of dilatation must be the origin if we are to use this matrix.

5   A square has its vertices at $A(1, 3)$, $B(4, 3)$, $C(4, 6)$ and $D(1, 6)$. Write down the matrix associated with the dilatation $[O, 2]$ and find the image of the square under this dilatation. Illustrate by a diagram.

6   Write down the matrices associated with the dilatations:

a $[O, 2]$ b $[O, -\frac{1}{2}]$ c $[O, -p]$

Illustrate by a diagram the effect of dilatations a and b on the square OABC where A is $(4, 0)$, B is $(4, 4)$ and C is $(0, 4)$.

7    $A(2, -1)$, $B(5, -1)$, $C(5, -3)$ and $D(2, -3)$ are vertices of a rectangle. Find the images of the vertices under the transformation whose matrix is $\begin{pmatrix} 2 & 0 \\ 0 & -2 \end{pmatrix}$.

Illustrate by a diagram, and describe the transformation as the composition of two geometrical transformations.

8   Repeat question 7 for the matrix $\begin{pmatrix} 0 & 3 \\ 3 & 0 \end{pmatrix}$.

(iii) Matrices associated with rotation

We have already met the matrices $\begin{pmatrix} 0 & -1 \\ 1 & 0 \end{pmatrix}$ and $\begin{pmatrix} 0 & 1 \\ -1 & 0 \end{pmatrix}$ which are associated with rotations of $\frac{1}{2}\pi$ and $-\frac{1}{2}\pi$ radians about the origin. These are particular cases of the more general transformation of rotation through θ radians about the origin.

In Figure 4, $P_1(x_1, y_1)$ is the image of $P(x, y)$ under a rotation about O of θ radians. Let $OP = r = OP_1$, and $\angle XOP = \alpha$ radians.

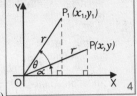

$x = r \cos \alpha$, and $y = r \sin \alpha$

$$x_1 = r \cos(\alpha + \theta) = r(\cos \alpha \cos \theta - \sin \alpha \sin \theta)$$
$$= (r \cos \alpha) \cos \theta - (r \sin \alpha) \sin \theta$$
$$= x \cos \theta - y \sin \theta$$

$$y_1 = r \sin(\alpha + \theta) = r(\sin \alpha \cos \theta + \cos \alpha \sin \theta)$$
$$= (r \cos \alpha) \sin \theta + (r \sin \alpha) \cos \theta$$
$$= x \sin \theta + y \cos \theta$$

Hence $\begin{pmatrix} x_1 \\ y_1 \end{pmatrix} = \begin{pmatrix} \cos \theta & -\sin \theta \\ \sin \theta & \cos \theta \end{pmatrix} \begin{pmatrix} x \\ y \end{pmatrix}$.

The matrix $\begin{pmatrix} \cos \theta & -\sin \theta \\ \sin \theta & \cos \theta \end{pmatrix}$ effects a rotation of θ radians about O.

If $\theta = \frac{1}{4}\pi$, the matrix is $\begin{pmatrix} \cos\frac{1}{4}\pi & -\sin\frac{1}{4}\pi \\ \sin\frac{1}{4}\pi & \cos\frac{1}{4}\pi \end{pmatrix}$, i.e. $\begin{pmatrix} \frac{1}{\sqrt{2}} & -\frac{1}{\sqrt{2}} \\ \frac{1}{\sqrt{2}} & \frac{1}{\sqrt{2}} \end{pmatrix}$.

If $\theta = -\frac{1}{2}\pi$, the matrix is $\begin{pmatrix} \cos(-\frac{1}{2}\pi) & -\sin(-\frac{1}{2}\pi) \\ \sin(-\frac{1}{2}\pi) & \cos(-\frac{1}{2}\pi) \end{pmatrix}$, i.e. $\begin{pmatrix} 0 & 1 \\ -1 & 0 \end{pmatrix}$.

Reminders

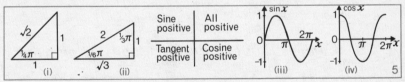

Exercise 3

1   Use suitable replacements for θ in $\begin{pmatrix} \cos \theta & -\sin \theta \\ \sin \theta & \cos \theta \end{pmatrix}$ to verify that the matrix associated with:

a   a half turn about the origin is $\begin{pmatrix} -1 & 0 \\ 0 & -1 \end{pmatrix}$.

b   a rotation of $\frac{1}{2}\pi$ radians about the origin is $\begin{pmatrix} 0 & -1 \\ 1 & 0 \end{pmatrix}$.

2 Write down and simplify the matrices giving these rotations about O:

   a $\frac{1}{6}\pi$ radians $\frac{1}{3}\pi$ radians $\frac{2}{3}\pi$ radians

3 A square has vertices A(2,0), B(0,2), C(−2,0), D(0,−2). Find the coordinates of the images of its vertices under a rotation about O of $\frac{1}{4}\pi$ radians. Illustrate by a sketch, noting that $2/\sqrt{2} = \sqrt{2} \doteqdot 1\cdot4$.

4 P is the point (4,0). If OP = OQ = OR, and ∠POQ = ∠QOR = ∠ROP, find the coordinates of Q and R.

5 The vertices of a triangle are A(0,2), B(−$\sqrt{3}$, −1) and C($\sqrt{3}$, −1). Find the images of the vertices under a rotation about O of $\frac{2}{3}\pi$ radians. Explain the result geometrically.

6 *a* Explain why the matrix $\begin{pmatrix} \frac{3}{5} & -\frac{4}{5} \\ \frac{4}{5} & \frac{3}{5} \end{pmatrix}$ is associated with a rotation about the origin of θ radians, where $\cos\theta = \frac{3}{5}$.

  b The vertex A of square OABC is (4,3). Find the coordinates of B and C, given that C lies in the second quadrant.

  c Find the images of the vertices under the rotation in *a*. Illustrate in a diagram.

7 *a* State the matrix associated with a rotation of $\frac{1}{3}\pi$ radians about O.

  b Calculate the images of A(2,0), B(1,$\sqrt{3}$), C(−1,$\sqrt{3}$), D(−2,0), E(−1, −$\sqrt{3}$) and F(1, −$\sqrt{3}$) under this rotation.

  c Explain the result geometrically.

---

2   *Composition of transformations by matrices*

We begin by investigating, both geometrically and by matrices, the composite transformation consisting of reflection in the *y*-axis (T_1), followed by rotation about O of $\frac{1}{2}\pi$ radians (T_2).

(i) *Geometrically*

Under T_1, P(a, b) → P$_1$(−a, b).
Under T_2, P$_1$(−a, b) → P$_2$(−b, −a).
Hence, in the usual notation,
under $T_2 \circ T_1$, P(a, b) → P$_2$(−b, −a).

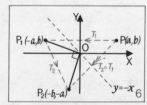

Since this is true for every point in the plane, we deduce that the composition of reflection in the y-axis, followed by rotation about O of $\frac{1}{2}\pi$ radians, is the transformation of reflection in the line $y = -x$.

(ii) *By matrices*

The matrices associated with T_1 and T_2 are $A = \begin{pmatrix} -1 & 0 \\ 0 & 1 \end{pmatrix}$ and $B = \begin{pmatrix} 0 & -1 \\ 1 & 0 \end{pmatrix}$.

For the coordinates of P_1, $\begin{pmatrix} -1 & 0 \\ 0 & 1 \end{pmatrix}\begin{pmatrix} a \\ b \end{pmatrix} = \begin{pmatrix} -a \\ b \end{pmatrix}$.

For the coordinates of P_2, $\begin{pmatrix} 0 & -1 \\ 1 & 0 \end{pmatrix}\begin{pmatrix} -a \\ b \end{pmatrix} = \begin{pmatrix} -b \\ -a \end{pmatrix}$.

But $\left[\begin{pmatrix} 0 & -1 \\ 1 & 0 \end{pmatrix}\begin{pmatrix} -1 & 0 \\ 0 & 1 \end{pmatrix}\right]\begin{pmatrix} a \\ b \end{pmatrix} = \begin{pmatrix} 0 & -1 \\ -1 & 0 \end{pmatrix}\begin{pmatrix} a \\ b \end{pmatrix} = \begin{pmatrix} -b \\ -a \end{pmatrix}$.

Hence BA is the matrix associated with the composite transformation $T_2 \circ T_1$. It is essential to multiply the matrices in the correct order, as BA is not usually equal to AB. $T_2 \circ T_1$ means 'T_1 *then* T_2'; in the product BA, the matrices are taken in corresponding order.

In general, since matrix multiplication is associative,

$$\begin{pmatrix} k & l \\ m & n \end{pmatrix}\left[\begin{pmatrix} p & q \\ r & s \end{pmatrix}\begin{pmatrix} a \\ b \end{pmatrix}\right] = \left[\begin{pmatrix} k & l \\ m & n \end{pmatrix}\begin{pmatrix} p & q \\ r & s \end{pmatrix}\right]\begin{pmatrix} a \\ b \end{pmatrix}.$$

Hence if we wish to operate *first* by $\begin{pmatrix} p & q \\ r & s \end{pmatrix}$, and *then* by $\begin{pmatrix} k & l \\ m & n \end{pmatrix}$, we can perform the composite transformation by using the product $\begin{pmatrix} k & l \\ m & n \end{pmatrix}\begin{pmatrix} p & q \\ r & s \end{pmatrix}$.

Exercise 4

1   Denote by X the operation of reflection in the x-axis, and by Y the operation of reflection in the y-axis.

a   What single transformation is equivalent to $Y \circ X$?

b   What is the image of $P(a, b)$ under this transformation?

c   Write down the matrices A and B associated with X and Y.

    d Simplify the matrix product BA, and then $(BA)\begin{pmatrix} a \\ b \end{pmatrix}$.

    e Verify that BA is the matrix associated with $Y \circ X$.

2   Denote by M_1 the operation of reflection in the line $y = x$, and by M_2 the operation of reflection in the line $y = -x$.

    a What single transformation is equivalent to $M_2 \circ M_1$?

    b What is the image of P(a, b) under this transformation?

    c Write down the matrices A and B associated with M_1 and M_2.

    d Verify that BA is the matrix associated with $M_2 \circ M_1$, and AB with $M_1 \circ M_2$. Why is $BA = AB$ in this special case?

3   Denote by X the operation of reflection in the x-axis, and by M the operation of reflection in the line $y = x$.

    a Give the single transformation equivalent to $M \circ X$, and write down the image of P(a, b).

    b Write down the matrices A and B associated with X and M, and verify that BA is the matrix associated with $M \circ X$.

    c Show that in this case $AB \neq BA$, and say what transformation is given by AB.

4   Find the matrix associated with reflection in the y-axis followed by a half turn about the origin. Interpret the result geometrically.

5 *a* Show by matrix multiplication that the operations of dilatation $[O, 2]$ (D) and reflection in the y-axis (Y) commute with each other, i.e. that $D \circ Y = Y \circ D$.

    b Noting that $\begin{pmatrix} k & 0 \\ 0 & k \end{pmatrix} = k\begin{pmatrix} 1 & 0 \\ 0 & 1 \end{pmatrix}$, explain why a dilatation always commutes with another 2×2 matrix transformation.

6 *a* What geometrical transformations are associated with the matrices $P = \begin{pmatrix} 1 & 0 \\ 0 & -1 \end{pmatrix}$ and $Q = \begin{pmatrix} -1 & 0 \\ 0 & 1 \end{pmatrix}$?

    b Show that $PQ = QP$: (*1*) geometrically (*2*) by matrix multiplication.

c If $I = \begin{pmatrix} 1 & 0 \\ 0 & 1 \end{pmatrix}$ and $H = PQ$,

copy and complete the composition table.

d Do all pairs of operations commute?

e Verify in a few cases the associative law for the operations.

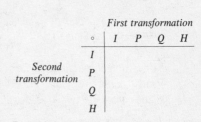

$\circ$	First transformation			
	I	P	Q	H
I				
Second transformation $\quad P$				
Q				
H				

7 Show that the matrix $\begin{pmatrix} 3 & -4 \\ 4 & 3 \end{pmatrix}$ effects the same transformation as the dilatation $[O, 5]$ followed by a rotation through an acute angle of θ radians about the origin where $\tan \theta = \frac{4}{3}$.

Do the transformations in this composition commute?

8 Let $R(\theta)$ represent the transformation of rotation through an angle θ radians about the origin. Use matrix multiplication to show that:

a $R(\theta_2) \circ R(\theta_1) = R(\theta_1 + \theta_2) = R(\theta_1) \circ R(\theta_2)$ b $R(\theta) \circ R(\theta) = R(2\theta)$.

9 In Figure 7, $OP = r$, $\angle XOP = \theta$ radians and $\angle XOA = \alpha$ radians.

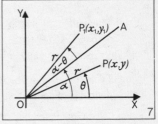

7

a Express x and y in terms of r and θ.

b If $P_1(x_1, y_1)$ is the image of P under reflection in OA, show that $x_1 = r \cos(2\alpha - \theta)$ and that $y_1 = r \sin(2\alpha - \theta)$.

c Deduce that $x_1 = x \cos 2\alpha + y \sin 2\alpha$ and that $y_1 = x \sin 2\alpha - y \cos 2\alpha$.

d Write down the matrix associated with the transformation which maps (x, y) to (x_1, y_1).

e What is the corresponding matrix associated with reflection in OB when $\angle XOB = \beta$ radians?

f Show by matrix multiplication that reflection in OA followed by reflection in OB gives a rotation of $2(\beta - \alpha)$ radians about O.

10 Under a composite mapping $P(a, b) \to P_1(a_1, b_1) \to P_2(a_2, b_2)$, where $a_1 = a + 2b$, $b_1 = a - 3b$, $a_2 = a_1$ and $b_2 = -b_1$.

a State the matrices associated with the two mappings, and with the composite mapping.

b Under this composite mapping find the image set of $\{(0, 0), (4, 0), (6, 3), (2, 3)\}$, and illustrate by a diagram.

11   The transformation $\begin{pmatrix} a_1 \\ b_1 \end{pmatrix} = \begin{pmatrix} \frac{4}{5} & -\frac{3}{5} \\ \frac{3}{5} & \frac{4}{5} \end{pmatrix} \begin{pmatrix} a \\ b \end{pmatrix}$ is followed by the transformation $\begin{pmatrix} a_2 \\ b_2 \end{pmatrix} = \begin{pmatrix} 5 & 0 \\ 0 & 5 \end{pmatrix} \begin{pmatrix} a_1 \\ b_1 \end{pmatrix}$.

a   State the geometrical transformations represented in each case.
b   Find the matrix which effects the combined transformation.
c   Check that the two transformations commute.

12a   If M is the matrix associated with a rotation of $\frac{1}{6}\pi$ radians about O, show by multiplication of matrices that M^2 is the matrix associated with a rotation of $\frac{1}{3}\pi$ radians about O.
b   Calculate M^3 as M^2M and as MM^2. Interpret M^3 geometrically.
c   Investigate the transformation whose matrix is M^6.

---

3   Composition of transformations geometrically

(i) Revision

The following is a list of reminders of the results arrived at in Book 7, Chapter 2, some of which we have already used in this chapter.

1   Two successive translations are equivalent to a translation: (i) by head-to-tail addition (ii) by addition of components (Figure 8).

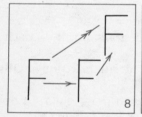 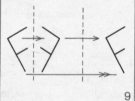 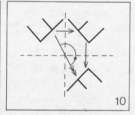

2   Two successive reflections in parallel axes are equivalent to a translation of: (i) magnitude twice the distance between the axes (ii) direction at right angles to the axes, from the first axis to the second (Figure 9).

3   Two successive reflections in perpendicular axes are equivalent to a half turn about the intersection of the axes (Figure 10).

4   Two successive rotations about the same centre are equivalent to a rotation about the same centre with magnitude equal to the sum of the two original magnitudes (Figure 11).

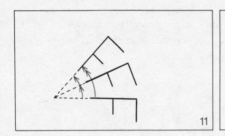

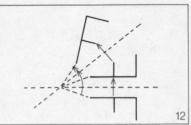

5   Two successive reflections in intersecting axes are equivalent to a rotation about the intersection of the axes of: (i) magnitude twice the angle between the axes (ii) sense from the first axis to the second (Figure 12).

6   (i) A transformation which maps a figure onto its image by sliding the plane is called a *direct* transformation (*D*).

(ii) A transformation which maps a figure onto its image by turning the plane over is called an opposite transformation (*O*).

Exercise 5

1   In Figure 13, G is the centroid of △ABC. (BD = DC, AG = 2GD.)
Find single representatives of the translations represented by:

a $\overrightarrow{AB}+\overrightarrow{BD}$ b $\frac{1}{2}\overrightarrow{BC}+\frac{1}{3}\overrightarrow{DA}$ c $\frac{2}{3}\overrightarrow{AD}+\overrightarrow{GC}$

d $\overrightarrow{BA}-\frac{2}{3}\overrightarrow{DA}$ e $\overrightarrow{CB}+\overrightarrow{BG}+\overrightarrow{GC}$ f $\overrightarrow{GC}-\frac{1}{3}\overrightarrow{AD}$

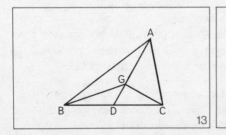

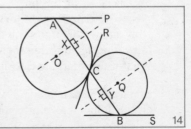

2 In Figure 14, two circles, centres O and Q, touch at C. ACB is a double chord. OX and QY are the perpendicular bisectors of AC and CB. Tangents AP, CR and BS are drawn at A, C and B.

a State the image of AP under reflection in OX.
b State the image of AP under successive reflections in OX and QY.
c What is the result of successive reflections in parallel axes?
d What can you conclude about AP and BS?

3 Find the values of p and q in each of the following.

a $\begin{pmatrix} 3 \\ 4 \end{pmatrix} + \begin{pmatrix} p \\ q \end{pmatrix} = \begin{pmatrix} -1 \\ 6 \end{pmatrix}$
b $\begin{pmatrix} p \\ 3 \end{pmatrix} - \begin{pmatrix} 4 \\ q \end{pmatrix} = \begin{pmatrix} 1 \\ 4 \end{pmatrix}$

c $\begin{pmatrix} 1 \\ 3 \end{pmatrix} + \begin{pmatrix} p \\ q \end{pmatrix} - \begin{pmatrix} 2 \\ -5 \end{pmatrix} = \begin{pmatrix} 10 \\ 8 \end{pmatrix}$
d $\begin{pmatrix} p \\ 2 \end{pmatrix} - \begin{pmatrix} 1 \\ q \end{pmatrix} + \begin{pmatrix} 2 \\ 3 \end{pmatrix} = \begin{pmatrix} q \\ p \end{pmatrix}$

4 a Find the image of $(2, -4)$ under successive reflections in $x = 3$ and $x = 7$.

b Find the image of $(-3, 2)$ under successive reflections in $y = -1$ and $y = 5$.

c If $(5, 1) \to (1, 1)$ under successive reflections in $x = 4$ and $x = h$, find h.

d Find the image of (a, b) under successive reflections in $x = h$ and $x = k$.

5 Let I represent the identity transformation, M_1 reflection in the line $y = x$, M_2 reflection in line $y = -x$, and H a half turn about the origin. Make up a composition table for I, M_1, M_2 and H. What are the inverses of M_1, M_2 and H?

6 a ABCD is a square S, with diagonals AC and BD drawn. What single transformation is equivalent to reflection in the line of AC followed by reflection in the line of AB? Sketch the image S_1 of S under successive reflections in these two lines.

b How many more times will the same composite transformation need to be repeated to bring the square back to its original position?

c What shape of figure is formed by BD and its images?

7 In the coordinate plane $\angle XOA = \frac{1}{8}\pi$ radians. What are the images of the following points under the composite transformation of reflection in the x-axis followed by reflection in OA?

a $(\sqrt{2}, -\sqrt{2})$ *b* $(2, 0)$ *c* $(0, 4)$

Find the images of the same points if the reflections are carried out in the reverse order.

8 *a* Copy and complete the composition table shown, where D represents a direct transformation and O represents an opposite transformation. What is the identity element?

$\circ$	D	O
D		
O		

b Make a 2×2 table for addition of even and odd numbers. Compare the tables.

(ii) Three or more transformations

Exercise 6

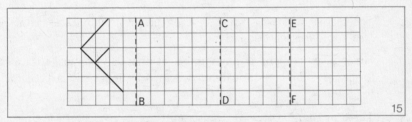

15

1   Copy Figure 15 on squared paper. If M_1, M_2 and M_3 denote the operations of reflection in the parallel axes AB, CD and EF respectively, sketch the images of F under:

a M_1 *b* $M_2 \circ M_1$ *c* $M_3 \circ M_2 \circ M_1$.

What single operation is equivalent to $M_2 \circ M_1$?
What single operation is equivalent to $M_3 \circ M_2 \circ M_1$?

2   Let M_1, M_2 and M_3 denote the operations of reflection in the lines with equations $x = 2$, $x = 3$ and $x = 7$ respectively.

a State the images of P(3, 2) under the transformations:

(*1*) M_1 (*2*) $M_2 \circ M_1$ (*3*) $M_3 \circ M_2 \circ M_1$.

How can you write down (*2*) without finding (*1*)?

b Find the image of Q (4, 4) under $M_3 \circ M_2 \circ M_1$.
c Find the image of R (a, b) under $M_3 \circ M_2 \circ M_1$.

3   Successive reflections in a set of parallel axes $\{l_1, l_2, ..., l_n\}$ can be replaced by a single transformation. What is the nature of the transformation for: *a* n even   *b* n odd?

4   Let M_1, M_2, M_3 and M_4 denote the operations of reflection in the x-axis, y-axis, the line $y = x$ and the line $y = -x$ respectively.

In a diagram on squared paper show the image of the triangle OAB with vertices O(0, 0), A(3, 0), B(3, 2) under the transformations:

   a M_1 b $M_2 \circ M_1$ c $M_3 \circ M_2 \circ M_1$ d $M_4 \circ M_3 \circ M_2 \circ M_1$

What conclusion is suggested?

5   A, B, C and D are the matrices associated with the operations M_1, M_2, M_3 and M_4 in question 4. Write down, and simplify where possible:

   a   A   b   BA   c   CBA   d   DCBA

What do you conclude?

6   n successive reflections in a set of n concurrent straight lines is either a rotation or a reflection. Explain.

7   Given a composition of any number of direct and opposite transformations, how could you decide whether the final result is direct or opposite?

8   Three or more successive translations are equivalent to a single transformation. What type of transformation? Define it.

9   Three or more successive rotations about a point are equivalent to a single transformation. What type of transformation? Define it.

10   Rotation through θ radians about a point followed by reflection in a line through the point are equivalent to a single transformation. Can you define this transformation?

A note on the glide reflection

Many more kinds of composition of transformations are possible. One particular type, consisting of the composition of a translation of the plane parallel to a given axis, followed by reflection in that axis, is called a *glide reflection*. The importance of the glide reflection is that if we include it with translation, rotation and reflection in an axis, we can map any given figure isometrically to any position in the plane. (An isometry is a transformation which preserves distance between points.)

If the transformation is direct, we can use translation or rotation; if it is opposite we use reflection or glide reflection.

a Figure 16(i) illustrates *translation*. In this case the image has its straight lines parallel to those in the original figure.

b If the lines in a direct image are not parallel we can effect the transformation by a *rotation*. Figure 16(ii) illustrates how the centre can be found by drawing two perpendicular bisectors, e.g. AO and BO.

c Figure 16(iii) illustrates the simple case of opposite transformation, *reflection* in an axis.

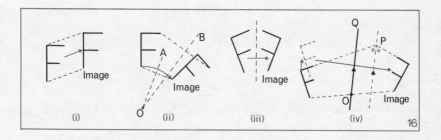

d In Figure 16(iv) no single transformation will do and we have to consider a *glide reflection*.

To find the necessary axis draw two corresponding lines meeting at P and bisect the angle between them.

Find O, the midpoint of the line joining two corresponding points and draw OQ parallel to the bisector.

Then a *translation* parallel to OQ *followed by reflection* in OQ will give the required image.

Since all of these transformations produce isometric figures, then *I* (identity), *T* (translation), *R* (rotation), *M* (reflection) and *G* (glide reflection) form a closed set for composition of transformations.

Sketch pairs of congruent shapes in different relative positions, and construct methods of mapping one member onto the other, as described above.

4 Inverse transformations

Since the product of a 2×2 matrix A and its inverse A^{-1} is the identity matrix I, the transformation whose matrix is A^{-1} must reverse, or 'undo', the transformation whose matrix is A.

Example 1. Find the matrices associated with a rotation about O of $\frac{1}{2}\pi$ radians, and with the inverse transformation.

We can find the image of point (x, y) under the given transformation as follows:

$$\begin{pmatrix} 0 & -1 \\ 1 & 0 \end{pmatrix}\begin{pmatrix} x \\ y \end{pmatrix} = \begin{pmatrix} -y \\ x \end{pmatrix}.$$

Remembering that the inverse of $\begin{pmatrix} a & b \\ c & d \end{pmatrix}$ is $\dfrac{1}{ad-bc}\begin{pmatrix} d & -b \\ -c & a \end{pmatrix}$, the inverse of $\begin{pmatrix} 0 & -1 \\ 1 & 0 \end{pmatrix}$ is $\dfrac{1}{0+1}\begin{pmatrix} 0 & 1 \\ -1 & 0 \end{pmatrix}$, i.e. $\begin{pmatrix} 0 & 1 \\ -1 & 0 \end{pmatrix}$, which we recognise as the matrix for a rotation about O of $-\frac{1}{2}\pi$ radians.

Also, $\begin{pmatrix} 0 & 1 \\ -1 & 0 \end{pmatrix}\begin{pmatrix} -y \\ x \end{pmatrix} = \begin{pmatrix} x \\ y \end{pmatrix}$, so by using the matrix and then its inverse we have $(x, y) \rightarrow (-y, x) \rightarrow (x, y)$.

Example 2. $\begin{pmatrix} 1 & 0 \\ 0 & 0 \end{pmatrix}\begin{pmatrix} x \\ y \end{pmatrix} = \begin{pmatrix} x \\ 0 \end{pmatrix}$; hence the transformation whose matrix is $\begin{pmatrix} 1 & 0 \\ 0 & 0 \end{pmatrix}$ maps every point in the plane to a point on the x-axis.

Since the determinant of this matrix is zero, there is no inverse and the transformation cannot be reversed.

This is obvious geometrically since each point on the x-axis is the image of a whole line of points in the plane and so no inverse transformation exists.

If a transformation whose matrix M maps the point P to P_1, then the inverse transformation whose matrix is M^{-1} (when M^{-1} exists) maps the point P_1 to P. The inverse of $\begin{pmatrix} a & b \\ c & d \end{pmatrix}$ is $\dfrac{1}{ad-bc}\begin{pmatrix} d & -b \\ -c & a \end{pmatrix}$.

Exercise 7

1 Describe the reflections with which the following matrices are associated.

$$a \begin{pmatrix} 1 & 0 \\ 0 & -1 \end{pmatrix} \quad b \begin{pmatrix} -1 & 0 \\ 0 & 1 \end{pmatrix} \quad c \begin{pmatrix} 0 & 1 \\ 1 & 0 \end{pmatrix} \quad d \begin{pmatrix} 0 & -1 \\ -1 & 0 \end{pmatrix}$$

Check that in each case the matrix is equal to its inverse. Explain the geometrical meaning.

2 a Write down the matrix associated with a rotation of θ radians about the origin.

  b Write down the inverse matrix.

  c What geometrical transformation is associated with the inverse?

3 a A(2, 1), B(5, 2) and C(3, 3) are the vertices of $\triangle$ABC. Find the coordinates of the images of these vertices under the transformation defined by $\begin{pmatrix} x_1 \\ y_1 \end{pmatrix} = \begin{pmatrix} 1 & -1 \\ 1 & 1 \end{pmatrix} \begin{pmatrix} x \\ y \end{pmatrix}$.

  b Find the matrix for the inverse transformation and check that it maps the image triangle to $\triangle$ABC.

4 The transformation whose matrix is $\begin{pmatrix} 2 & 1 \\ -1 & 1 \end{pmatrix}$ maps parallelogram PQRS onto parallelogram $P_1Q_1R_1S_1$ whose vertices are $P_1(3, 0)$, $Q_1(14, -4)$, $R_1(18, -3)$ and $S_1(7, 1)$.

  Use the inverse matrix to find the coordinates of P, Q, R and S.

5 a A transformation maps (2, 1) to (4, 3) and (3, 5) to (13, 1). Find the matrix $\begin{pmatrix} a & b \\ c & d \end{pmatrix}$ of the transformation.

  b Obtain the matrix of the inverse transformation and check by using the matrix on the points (4, 3) and (13, 1).

6 a What geometrical transformation is associated with the matrix $\begin{pmatrix} 1 & 1 \\ 0 & 0 \end{pmatrix}$?

  b Does this matrix have an inverse?

  c Explain (1) geometrically (2) algebraically why the mapping associated with this matrix cannot be 'undone'.

7 *a*   Which matrix will effect a dilatation $[O, 3]$?

  b   Which matrix is associated with the inverse of this dilatation?

8   Let $A = \begin{pmatrix} 1 & 0 \\ 0 & -1 \end{pmatrix}$ and $B = \begin{pmatrix} 0 & 1 \\ 1 & 0 \end{pmatrix}$.

  a   State the transformations effected by A, B, BA and AB.

  b   Find out if $B^{-1}A^{-1}$ or $A^{-1}B^{-1}$ is the inverse of BA.

  c   Explain your answer to *b* geometrically.

9 *a*   If A and B are 2×2 matrices whose inverses exist, use the associative law to simplify $(B^{-1}A^{-1})(AB)$.

  b   What is the inverse of AB? (Note that we could think of this as a 'socks and shoes' rule. We put on our socks and then our shoes. To reverse the process we take off our shoes and then our socks.)

10   Transformations are defined by the equations:

$$x_1 = 2x - 3y, \; y_1 = x + 2y, \; x_2 = 3x_1 + y_1 \text{ and } y_2 = x_1 - y_1.$$

    Find the matrices which will effect the transformations:

  a $(x, y) \rightarrow (x_2, y_2)$ *b* $(x_2, y_2) \rightarrow (x, y)$

11*a*   State the geometrical transformations associated with the matrices $P = \begin{pmatrix} -1 & 0 \\ 0 & 1 \end{pmatrix}$ and $Q = \begin{pmatrix} 0 & 1 \\ -1 & 0 \end{pmatrix}$.

  b   If $T = PQ$, calculate T. State the geometrical transformation with which T is associated.

  c   Repeat *b* for $S = QP$.

  d   What are the inverse matrices T^{-1} and S^{-1}? Explain your answer geometrically.

---

5   *Transformation of loci*

Example. Find the image of the line $x + 3y + 2 = 0$ under the transformation associated with the matrix $\begin{pmatrix} 2 & 3 \\ 1 & 2 \end{pmatrix}$.

If (a, b) lies on the line $x + 3y + 2 = 0$, then $a + 3b + 2 = 0$. If (a_1, b_1) is the image of (a, b) then

$$\begin{pmatrix} a_1 \\ b_1 \end{pmatrix} = \begin{pmatrix} 2 & 3 \\ 1 & 2 \end{pmatrix} \begin{pmatrix} a \\ b \end{pmatrix}.$$

Premultiply both sides by the inverse matrix $\begin{pmatrix} 2 & -3 \\ -1 & 2 \end{pmatrix}$, giving

$$\begin{pmatrix} a \\ b \end{pmatrix} = \begin{pmatrix} 2 & -3 \\ -1 & 2 \end{pmatrix} \begin{pmatrix} a_1 \\ b_1 \end{pmatrix} = \begin{pmatrix} 2a_1 - 3b_1 \\ -a_1 + 2b_1 \end{pmatrix} \text{ i.e. } \begin{aligned} a &= 2a_1 - 3b_1 \\ b &= -a_1 + 2b_1 \end{aligned}$$

Since $a + 3b + 2 = 0$, then

$$(2a_1 - 3b_1) + 3(-a_1 + 2b_1) + 2 = 0$$

$$\Leftrightarrow \quad -a_1 + 3b_1 + 2 = 0.$$

Hence (a_1, b_1) must satisfy the equation $-x + 3y + 2 = 0$, which is the equation of the image line.

This method is not available when the matrix under investigation has no inverse. The image in such a case is usually obvious, see Example 2, page 130.

Examples involving composite transformations may be reduced to examples like those above by multiplication of matrices.

Note. Under transformations associated with 2×2 matrices the image of a straight line must always be a straight line.

If $\quad \begin{pmatrix} a_1 \\ b_1 \end{pmatrix} = \begin{pmatrix} p & q \\ r & s \end{pmatrix} \begin{pmatrix} a \\ b \end{pmatrix} \quad$ then $\quad \begin{pmatrix} a \\ b \end{pmatrix} = \dfrac{1}{ps - qr} \begin{pmatrix} s & -q \\ -r & p \end{pmatrix} \begin{pmatrix} a_1 \\ b_1 \end{pmatrix}.$

Hence the replacements for a and b are always linear expressions in a_1 and b_1. It follows that the images of straight-sided figures are straight-sided figures, and we need concern ourselves only with the images of the vertices. This has been taken for granted in some of the earlier examples in this chapter.

Exercise 8

1 a Under the transformation associated with the matrix $A = \begin{pmatrix} 1 & 1 \\ 0 & 1 \end{pmatrix}$, point $(a, b) \to$ point (a_1, b_1). Express this as a matrix equation.

b Use the inverse matrix A^{-1} to express a and b in terms of a_1 and b_1.

c Hence find the equation of the image of the line $x+y+1=0$ under the transformation associated with A. Illustrate with a diagram.

2   In each of the following cases find the equation of the image of the given line under the transformation associated with the given matrix.

a $\begin{pmatrix} 1 & 0 \\ 0 & -1 \end{pmatrix}$, $y = 2x$ *b* $\begin{pmatrix} 1 & 2 \\ 0 & 1 \end{pmatrix}$, $y = x+1$

c $\begin{pmatrix} 2 & 3 \\ 1 & 2 \end{pmatrix}$, $2x+3y+1 = 0$ *d* $\begin{pmatrix} 1 & -3 \\ 2 & -5 \end{pmatrix}$, $x-2y+3 = 0$

3   Find the image of the line $2x-3y+4=0$ under:

a reflection in the y-axis   *b* a rotation of $-\frac{1}{2}\pi$ radians about O

c a transformation where $(a,b) \to (a_1,b_1)$ with $a_1 = 5a+2b$ and $b_1 = 4a+2b$.

4   Use matrix multiplication to find the image of the line $3x-y+2=0$ under:

a reflection in the line $y = x$ followed by a rotation of $\frac{1}{2}\pi$ radians about O

b reflection in the x-axis followed by a transformation $(a, b) \to (a_1, b_1)$ such that $a_1 = a+b$ and $b_1 = b$.

5   Let $A = \begin{pmatrix} 1 & 1 \\ 1 & -1 \end{pmatrix}$, $B = \begin{pmatrix} a & b \\ 1 & -1 \end{pmatrix}$ and $BA = \begin{pmatrix} 4 & 2 \\ 0 & 2 \end{pmatrix}$.

a Calculate the values of a and b.

b Calculate AB and show that it represents the transformation associated with $\begin{pmatrix} 2 & 0 \\ 1 & 1 \end{pmatrix}$ followed by a dilatation $[O, 2]$.

c Find the equation of the image of the line $y = 2x-1$ under the transformation associated with AB.

6   Find the equations of the images of the following straight lines under a rotation of $\frac{1}{4}\pi$ radians about the origin.

a $2x-y = 3$ $x+y+2 = 0$

7 *a* Find the image of the circle $x^2 + y^2 = 9$ under the transformation associated with the matrix $\begin{pmatrix} 1 & 1 \\ 0 & 1 \end{pmatrix}$.

  b Find the images of the four points $(\pm 3, 0)$, $(0, \pm 3)$ which lie on the circle.

  c Make a rough sketch of the image.

8 *a* Find the image of the circle $x^2 + y^2 - 6x + 8y + 9 = 0$ under reflection in the line $y = -x$.

  b What are the centre and radius of the given circle?

  c What are the centre and radius of the image? Illustrate by a sketch.

9 Show that the circle $x^2 + y^2 = 16$ maps onto itself under the transformation associated with the matrix $\begin{pmatrix} \cos\theta & -\sin\theta \\ \sin\theta & \cos\theta \end{pmatrix}$.

10*a* Find the image of the parabola $y^2 = 2x$ under the reflection associated with the matrix $\begin{pmatrix} -1 & 0 \\ 0 & 1 \end{pmatrix}$.

  b Find the image after a further transformation of a rotation of $\frac{1}{4}\pi$ radians about the origin. Why must this final image still be a parabola?

11 Find the image of the parabola $y = 2x^2 + 1$ under the dilatation $[O, 3]$.

12 Find the equation of the image of the hyperbola $x^2 - y^2 = 4$ under a rotation about O of $\frac{1}{4}\pi$ radians.

13*a* Show that the image of the circle $x^2 + y^2 = 1$ under the mapping whose matrix is $\begin{pmatrix} 2 & 0 \\ 0 & 2 \end{pmatrix}$ is a concentric circle. Illustrate by a diagram.

  b Repeat *a* for the mapping whose matrix is $\begin{pmatrix} 2 & 0 \\ 0 & 1 \end{pmatrix}$, using the same diagram.

6 Some other matrix transformations and properties

(i) The shear and other transformations

The transformations we have studied in this chapter are special cases of the transformation defined by the equations

$$\begin{matrix} x_1 = ax + by \\ y_1 = cx + dy \end{matrix} \quad \text{i.e.} \quad \begin{pmatrix} x_1 \\ y_1 \end{pmatrix} = \begin{pmatrix} a & b \\ c & d \end{pmatrix} \begin{pmatrix} x \\ y \end{pmatrix}$$

where a, b, c and d are real numbers.

Since these equations are linear the mapping is called a *linear transformation*. The origin $(0, 0)$ maps to $(0, 0)$ in every case since $\begin{pmatrix} a & b \\ c & d \end{pmatrix}\begin{pmatrix} 0 \\ 0 \end{pmatrix} = \begin{pmatrix} 0 \\ 0 \end{pmatrix}$, i.e. the origin is an invariant point. We saw in Section 5 that under every linear transformation straight lines are transformed to straight lines.

Apart from translation, all of the transformations in this Section are of the linear type:

e.g. for reflection in the line $y = x$, $a = d = 0$, $b = c = 1$;

for rotation of θ radians about the origin, $a = d = \cos\theta$, $b = -c = -\sin\theta$;

for dilatation with centre at the origin, $a = d$, $b = c = 0$.

Example 1. Consider the transformation for which $a = d = 1$, $b = 2$ and $c = 0$ acting on the square whose vertices are $O(0, 0)$, $A(2, 0)$, $B(2, 2)$ and $C(0, 2)$.

$$\begin{pmatrix} 1 & 2 \\ 0 & 1 \end{pmatrix}\begin{pmatrix} 0 & 2 & 2 & 0 \\ 0 & 0 & 2 & 2 \end{pmatrix} = \begin{pmatrix} 0 & 2 & 6 & 4 \\ 0 & 0 & 2 & 2 \end{pmatrix}$$

The images are $O(0, 0)$, $A_1(2, 0)$, $B_1(6, 2)$ and $C_1(4, 2)$ as shown in Figure 17. Each point on the x-axis is its own image but all other points move parallel to the x-axis through a distance proportional to their distance from that axis. Such a transformation is called a *shear* in the direction of the x-axis. The square OABC has been sheared to the parallelogram $OA_1B_1C_1$ of the same area. You can get the same effect with a pack of playing cards (see Figure 18).

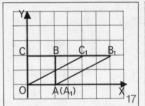
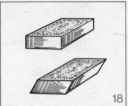
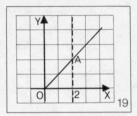

Example 2. Explain geometrically the transformation effected by the matrix $\begin{pmatrix} 1 & 0 \\ 1 & 0 \end{pmatrix}$.

Since $\begin{pmatrix} 1 & 0 \\ 1 & 0 \end{pmatrix}\begin{pmatrix} x \\ y \end{pmatrix} = \begin{pmatrix} x \\ x \end{pmatrix}$, every point in the plane is transformed to a point with equal coordinates.

Hence the whole plane is transformed to the straight line $y = x$ (see Figure 19). The point A is the image of the whole straight line $x = 2$.

Note. Since in this case there is not a one-to-one correspondence between points in the plane, it is not possible to reverse the transformation, e.g. given a point $(3, 3)$ on the line $y = x$, we cannot find a unique point in the plane corresponding to it. It is worth noting that the matrix $\begin{pmatrix} 1 & 0 \\ 1 & 0 \end{pmatrix}$ has no inverse.

Exercise 9B

1 The vertices of a square are A$(2, 1)$, B$(3, 1)$, C$(3, 2)$ and D$(2, 2)$. Find the images of these points under the transformation whose matrix is $\begin{pmatrix} 1 & 1 \\ 0 & 1 \end{pmatrix}$.

   Illustrate by a diagram. Describe the effect of the transformation on the shape and area of the square.

2 Repeat question *1* for the rectangle with vertices P$(1, 0)$, Q$(4, 0)$, R$(4, 2)$ and S$(1, 2)$.

3 Find the image of the square ABCD of question *1* by using the matrix $\begin{pmatrix} 2 & 3 \\ 0 & 2 \end{pmatrix}$. Describe the effect of this transformation.

4   Repeat question *3* for the rectangle PQRS of question *2*.

5   Investigate the effect of the shear whose matrix is $\begin{pmatrix} 1 & 0 \\ 2 & 1 \end{pmatrix}$ on the square ABCD (question *1*) and the rectangle PQRS (question *2*). Illustrate by diagrams.

6   If $(x, y) \rightarrow (x_1, y_1)$, where $x_1 = 3x$ and $y_1 = y$, state the matrix associated with the transformation.

      Use this matrix to transform the square OABC where A is $(1, 0)$, B is $(1, 1)$ and C is $(0, 1)$.

      What is the geometrical result of this transformation?

7   Find the image of the set $\{(1, 4), (0, 2), (-2, 4), (a, b)\}$ under the transformation associated with each of the matrices:

      a $\begin{pmatrix} 4 & 0 \\ 1 & 0 \end{pmatrix}$ b $\begin{pmatrix} 1 & 1 \\ 0 & 0 \end{pmatrix}$ c $\begin{pmatrix} 0 & 0 \\ 4 & 0 \end{pmatrix}$ d $\begin{pmatrix} 0 & 0 \\ 0 & 0 \end{pmatrix}$.

      What conclusions do you draw in each case?

8   Why cannot the transformations in question *7* be reversed? Give an algebraic reason as well as a geometrical one.

9   Indicate in a sketch the images of the square OABC where A is $(2, 0)$, B is $(2, 2)$ and C is $(0, 2)$ under the transformations whose matrices are: a $\begin{pmatrix} 1 & 2 \\ 3 & 1 \end{pmatrix}$ b $\begin{pmatrix} 2 & 1 \\ 4 & 2 \end{pmatrix}$.

10   The vertices of a parallelogram are $A(1, 1)$, $B(4, 2)$, $C(5, 4)$ and $D(2, 3)$. Find the image of ABCD under the transformation whose matrix is $\begin{pmatrix} 3 & 1 \\ 1 & 1 \end{pmatrix}$. Illustrate by a diagram.

(ii) Geometrical meaning of the determinant of a matrix

In Worked Example 1 above we found that the transformation whose matrix was $\begin{pmatrix} 1 & 2 \\ 0 & 1 \end{pmatrix}$ transformed a square into a parallelogram of the *same* area. In this case the determinant of the matrix $= (1 \times 1) - (2 \times 0) = 1$.

If a unit square is transformed by the dilatation matrix $\begin{pmatrix} 3 & 0 \\ 0 & 3 \end{pmatrix}$, then it becomes a square on a side of 3 units and its area is 9 square units. In this case the determinant of the matrix

$$= (3 \times 3) - (0 \times 0) = 9.$$

See also questions *1–5* of Exercise 9B.

Exercise 10B

In questions *1–4* use diagrams on squared paper to find the area of the given figure and the area of the image. Calculate also the value of the determinant of the matrix. (Use circumscribing rectangles, if necessary, to calculate areas.)

	Vertices of given figure	*Transformation matrix*
1	O (0,0), A (1,0), B (1,1), C (0,1)	$\begin{pmatrix} 2 & 3 \\ 0 & 2 \end{pmatrix}$
2	P (2,1), Q (5,1), R (5,3), S (2,3)	$\begin{pmatrix} 0 & -1 \\ 1 & 0 \end{pmatrix}$
3	A (2,1), B (6,1), C (6,4)	$\begin{pmatrix} 2 & 0 \\ 0 & -2 \end{pmatrix}$
4	L (−1,0), M (1,0), N (0,4)	$\begin{pmatrix} 4 & -3 \\ 3 & 4 \end{pmatrix}$

In each case that we have considered, the area of the image was $|ad - bc| \times$ the original area, where $\begin{pmatrix} a & b \\ c & d \end{pmatrix}$ is the transformation matrix, and $|ad - bc|$ denotes the numerical value of $ad - bc$. $|ad - bc|$ determines the *area magnification*.

The general proof of this theorem is outside the scope of an elementary treatment of transformations.

5   Check that the theorem about areas is true for the unit square OABC with vertices O(0,0), A(1,0), B(1,1) and C(0,1) under transformations whose matrices are:

a $\begin{pmatrix} 0 & 2 \\ -2 & 0 \end{pmatrix}$ b $\begin{pmatrix} 3 & 1 \\ 1 & 2 \end{pmatrix}$ c $\begin{pmatrix} 4 & 1 \\ 1 & 1 \end{pmatrix}$ d $\begin{pmatrix} 3 & 4 \\ 0 & 3 \end{pmatrix}$.

6   By a consideration of area magnification, state what will happen to the vertices of a square under the transformation whose matrix is $\begin{pmatrix} 2 & 4 \\ 1 & 2 \end{pmatrix}$.

Summary

1

Transformation	Matrix	Transformation	Matrix
Identity	$\begin{pmatrix} 1 & 0 \\ 0 & 1 \end{pmatrix}$	Reflection in $y = x$	$\begin{pmatrix} 0 & 1 \\ 1 & 0 \end{pmatrix}$
Half turn about O	$\begin{pmatrix} -1 & 0 \\ 0 & -1 \end{pmatrix}$	Reflection in $y = -x$	$\begin{pmatrix} 0 & -1 \\ -1 & 0 \end{pmatrix}$
Reflection in x-axis	$\begin{pmatrix} 1 & 0 \\ 0 & -1 \end{pmatrix}$	Rotation about O of $\frac{1}{2}\pi$ radians	$\begin{pmatrix} 0 & -1 \\ 1 & 0 \end{pmatrix}$
Reflection in y-axis	$\begin{pmatrix} -1 & 0 \\ 0 & 1 \end{pmatrix}$	Rotation about O of $-\frac{1}{2}\pi$ radians	$\begin{pmatrix} 0 & 1 \\ -1 & 0 \end{pmatrix}$

. . . .

Dilatation $[O, k]$	$\begin{pmatrix} k & 0 \\ 0 & k \end{pmatrix}$	Rotation about O of θ radians	$\begin{pmatrix} \cos\theta & -\sin\theta \\ \sin\theta & \cos\theta \end{pmatrix}$

2   The composition of transformations associated with 2×2 matrices can be effected by multiplication of matrices in the appropriate order.

3   Two successive translations are equivalent to a translation.

4   Two successive reflections in parallel axes are equivalent to a translation.

5   Two successive rotations about the same centre are equivalent to a rotation about that centre.

6   Two successive reflections in intersecting axes are equivalent to a rotation about the point of intersection of the axes.

7   Three or more transformations can be combined geometrically or using matrices.

8   If a transformation whose matrix is A maps P to P_1, and the inverse matrix A^{-1} exists, then the inverse transformation whose matrix is A^{-1} maps P_1 to P.

9   If a matrix A maps (a, b) to (a_1, b_1), equations of loci may be transformed by the use of $\begin{pmatrix} a \\ b \end{pmatrix} = A^{-1}\begin{pmatrix} a_1 \\ b_1 \end{pmatrix}$, if A^{-1} exists.

Mathematical Deduction and Proof

3

1 Open sentences and statements

Suppose that we are asked to decide whether each of the following is true or false:

   (i) $x - 5 = -1$.

   (ii) It is a cylindrical object.

   (iii) All metals are good conductors of electricity.

   (iv) -7 is greater than 5.

We cannot decide in the case of (i) and (ii) until we are given replacements for 'x' and 'it'; these are *open sentences*.

But we can say that (iii) is true and that (iv) is false; these sentences are called statements. An open sentence with a variable x gives rise, for each replacement of x, to a statement, and this statement is either true or false.

Exercise 1

Which of the following are statements? Which of the statements are true, and which are false?

1   There are twelve months in a year.

2   Soopa is a better buy!

3   111 is a prime number.

4    $-5(3 - 7) = 20$.

5    $x + 3 = -1$.

6    $\pi^2 = 10$.

7    $2n + 1$ is an odd number, $n \in N$.

8    $(-5)*(-3) = 15$.

9   For every integer y, $y < 2y$.

10   a is a factor of 12.

11   The tenth term of the sequence of squares 1, 4, 9, 16, ..., is 100.

12   $\text{Cos}^2 a° + \sin^2 a° = 1$, for all $a \in R$.

13   The product of two $m \times n$ matrices ($m \neq n$) is a square matrix.

14   The product of the gradients of two perpendicular lines (excluding $x = 0$ and $y = 0$ and lines parallel to the axes) is -1.

15   In every $\triangle$ABC: a $a+b = c$ b $a+b > c$ c $a^2+b^2 = c^2$

16   Every geometric series has a sum to infinity.

17   The scalar product of two vectors is a real number.

18   $(a+b+c)^2 = a^2 + b^2 + c^2 + 2ab + 2bc + 2ca$ $(a, b, c \in R)$.

19   The derivative of x^n is nx^{n-1}.

20   The line $y = mx + c$ is a tangent to the circle $x^2 + y^2 = r^2$.

---

2   *Implication*

Sentences may be combined to form compound sentences by using connecting words such as *and*, *or*, *but*. Of particular interest are compound sentences of the form '*If* . . . , *then* . . .' which occur frequently in mathematics as well as in our daily conversation.

Example 1. *If* John gets up after 9 am, *then* he will be late for school.

Example 2. *If* $x > 5$, then $2x > 10$.

Notice that in the above examples, the dotted parts in 'If . . . , then . . .' are replaced by sentences such that when the first sentence is true, the second sentence is true also. We say that the first sentence *implies* the second. For example, we can say: 'John gets up after 9 am' implies that 'he will be late for school'. This kind of 'if . . . then . . .' statement is called an *implication*.

Let p and q be variables which are replaceable by sentences. An *implication* is a sentence of the form 'If p, then q', or symbolically, $p \Rightarrow q$ where the symbol $\Rightarrow$ represents the word 'implies'.

For instance, let p be replaced by '$x > 5$' and q by '$2x > 10$'. 'If p, then q' means: 'If $x > 5$, then $2x > 10$' or '$x > 5 \Rightarrow 2x > 10$'.

Note. For an implication $p \Rightarrow q$ *which is true*, we cannot have p true and q false. If p is true and q is false, then $p \Rightarrow q$ is *false*.

Exercise 2

Copy and complete the following to give true implications:

1   If a number is greater than 10, its square is ...

2   If $3n + 5 = 2$ and $n \in Z$, then $n = \ldots$

3   If the point $(3, 5)$ is reflected in the origin, then its image is the point $(\ldots, \ldots)$.

4   $A = \pi r^2 \Rightarrow r = \ldots$

5   If $\frac{1}{2}\pi < x < \pi$, then $\cos x$ is a ... number.

6   Given $0 < x < \frac{1}{2}\pi$, $\tan x = 1 \Rightarrow x = \ldots$

7   $y = 6x^2 \Rightarrow \displaystyle\int_{-1}^{1} y \, dx = \ldots$

8   If A is the point (x_1, y_1) and B is (x_2, y_2), then the length of AB $= \ldots$

Assuming that the variables are on the set of *real numbers*, which of the following implications are true and which are false?

9   $x = 5 \Rightarrow x^2 = 25$

10   $x^2 = 25 \Rightarrow x = 5$

11   $a < 0 \Rightarrow a^2 > 0$

12   $\frac{1}{4} < \frac{1}{2} \Rightarrow \log\frac{1}{4} < \log\frac{1}{2}$

13   $ax^2 + bx + c = 0$ has real roots $\Rightarrow b^2 - 4ac < 0$.

14   $f(h) = 0 \Rightarrow x - h$ is a factor of polynomial $f(x)$.

15   $f'(a) = 0 \Rightarrow f(a)$ is a maximum value of f.

16   $\boldsymbol{u} \cdot \boldsymbol{v} = 0 \Rightarrow \boldsymbol{u}$ is perpendicular to $\boldsymbol{v}$.

17   If $X = $ reflection in the x-axis, $Y = $ reflection in the y-axis, and $H = $ half turn about O, then $X \circ Y = H$.

18   If $g(x) = x^2$ and $f(x) = x-1$, then $g(f(x)) = x^2-1$

Construct an implication from each of the following:

19   $8 = 2^3, \log_2 8 = 3$ 20 $f(x) = x^2 - \dfrac{1}{x^2}, f(-2) = \ldots$

21   $y = \sin^2 x, \dfrac{dy}{dx} = \ldots$ 22 $x = \tfrac{3}{2}\pi, \sin x = \ldots$

3   The converse of an implication

Here are two pairs of implications:

1 (i) $a > 0 \;\Rightarrow\; a^3 > 0$ (ii) $a^3 > 0 \;\Rightarrow\; a > 0$

2 (i) $a = 0 \;\Rightarrow\; ab = 0$ (ii) $ab = 0 \;\Rightarrow\; a = 0$

In each pair, the sentences in (i) have been interchanged to give a new implication (ii) which is called the *converse* of (i).

The examples show that the converse of a true implication may, or may not, be true. The converse 1(ii) is true but the converse 2(ii) is false, since b could be zero in the consequent clause. To sum up,

Implication	Converse
$p \Rightarrow q$	$q \Rightarrow p$

An implication can sometimes be shown to be false by citing a *single* example, called a *counter-example*, for which it is false.

Example. State the converse of '*If x is a multiple of* 9, *then x is a multiple of* 3', and give a counter-example to prove that the converse is false.

Converse: If x is a multiple of 3, then x is a multiple of 9.

Counter-example: Replace x by 15. 'If 15 is a multiple of 3, then 15 is a multiple of 9' is a false implication. Hence the converse is false, no matter how many true examples can be found.

Exercise 3

State in words the converse of each of the following.

1 If a number ends in zero, it is divisible by 5.

2 If a man is a guardsman, he is over 180 cm tall.

3 If n is a prime number greater than 2, then n is an odd number.

4 If a quadrilateral is a square, its diagonals intersect at right angles.

5 In a triangle ABC, if A is a right angle, then B and C are both acute angles (or $A = 90° \Rightarrow B < 90°$ and $C < 90°$).

6 Suppose that $x \in Z$. $x = 3 \Rightarrow x^2 = 9$.

7 If a plane figure has two perpendicular axes of symmetry, then it has a centre of symmetry.

8 If a and b are odd numbers, then $a + b$ is even.

9 If 3 is a root of $x^2 + x - k = 0$, k is a multiple of 3.

10 Suppose that a line L is the perpendicular bisector of line AB. 'If P is on the same side of L as B, then $PA \neq PB$.

11 Now show by giving just one counter-example-in each case, that every one of the converses you have stated in questions *1* to *10* is false.

State the converse of each of the following theorems. Then say whether each theorem and each converse is true or false.

12 If two triangles are congruent, then their areas are equal.

13 'If two triangles have three angles of one respectively equal to three angles of the other, then the triangles are congruent.

14 If a quadrilateral is cyclic, its opposite angles are supplementary.

15 If two angles are in the same segment of a circle, the angles are equal.

4 *Two-way implications – equivalence*

Figure 1 shows reflection in an axis XY, with P and P′ corresponding points. It follows at once that:

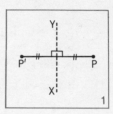

P′ is the image of P

⇒ P′P is bisected at right angles by XY.

The *converse* of this implication is:

P′P is bisected at right angles by XY ⇒ P′ is the image of P.

Since the converse is also true, we have a *two-way implication*, or *equivalence*, which is stated as follows:

P′ is the image of P ⇔ P′P is bisected at right angles by XY.

The equivalence symbol ⇔ can be read in either of two ways:

(i) *is equivalent to* (ii) *if and only if.*

‘$p ⇔ q$’ means that $p ⇒ q$ and $q ⇒ p$ are both true. It can be read as ‘p is equivalent to q’ or ‘p if and only if q’.

You have already met a number of true implications, or ‘theorems’, in geometry which have true converses. Some of these are included in Exercise 4.

Exercise 4

In questions *1* to *10* state whether the implication can be replaced by a two-way implication; if it can, make the replacement using the equivalence sign.

1   If $a = b$, then $a+c = b+c$.   2   If $A ⊂ B$, then $A ∩ B = A$.

3   $x = y ⇒ -x = -y$.   4 $n = -3 ⇒ n^2 = 9, n ∈ R$.

5   $x ∈ A ∩ B ⇒ x ∈ A$.   6   If n is odd, then n^2 is odd, $n ∈ Z$.

7   For rectangular axes, the straight line L slopes up from left to right ⇒ the gradient of L is positive.

8 If PQ and PR are tangents to a circle centre O, then OQPR is a kite.

9 Triangle ABC is right-angled at A $\Rightarrow a^2 = b^2 + c^2$.

10 If $y = 1 - x^2$, then $\dfrac{dy}{dx} = -2x$.

In questions *11* to *14*, state the converse of each implication and say whether the converse is true or false. If true, write out the corresponding two-way implication.

11 If two triangles are equiangular, their corresponding sides are in proportion.

12 If two chords of a circle are equal, they are equidistant from the centre.

13 If AB is parallel to A'B' and AC is parallel to A'C', then angle BAC = angle B'A'C'.

14 If PQ is equal and parallel to RS and in the same direction, PQSR is a parallelogram.

In questions *15* to *18*, construct, where possible, two-way implications.

15 $A \subset B$, and $A \cup B = B$.   16 $x > 3$, and $x^3 > 30$.

17 'A diameter of a circle is perpendicular to a chord', and 'a diameter of a circle bisects a chord'.

18 'A triangle has two axes of bilateral symmetry', and 'a triangle is equilateral'.

---

5 *Quantified statements*

Consider the following statements which are assumed to be true:
   (i) All cats have tails.
   (ii) Some pilots are women.
   (iii) No even number has a square which is odd.

These statements are concerned with sets of objects, and involve the use of the quantifiers *all* (*every*, *each*), *some* (*at least one*) or *no*.

Quantifiers indicate the notion of 'how many'. Statements such as (i), (ii) and (iii) are called *quantified statements* and we now examine these in detail.

Example 1. '*All cats have tails.*'

Let $E = \{$all animals$\}$, $C = \{$cats$\}$, and $T = \{$animals with tails$\}$. Translating the given statement into a statement about sets, $C \subset T$. The Venn diagram in Figure 2 illustrates this information.

Note that the statements 'Each cat has a tail' and 'Every cat has a tail' are each equivalent to the given statement 'All cats have tails'.

Note also that each of these quantified statements is equivalent to the implication, 'If x is a cat, then x has a tail,' and the statement about sets is equivalent to '$x \in C \implies x \in T$'.

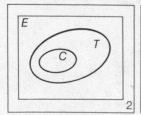 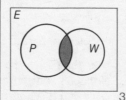 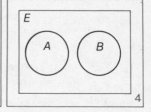

Example 2. '*Some pilots are women.*'

Let $E = \{$all people$\}$, $P = \{$pilots$\}$, and $W = \{$women$\}$. Translating the statement into a statement about sets, $P \cap W \neq \phi$. The Venn diagram in Figure 3 illustrates this information.

Note that the word '*some*' means '*one or more*', and does not exclude the possibility of *all*. An equivalent statement would be 'At least one pilot is a woman'. Can we deduce that 'some women are pilots'?

Example 3. '*No even number has a square which is odd.*'

Let $E = \{$all integers$\}$, $A = \{$even numbers$\}$, and $B = \{$integers with odd squares$\}$. Translating the statement into a statement about sets, $A \cap B = \phi$. The Venn diagram in Figure 4 illustrates this information.

Exercise 5

Using the letters suggested, translate each of the following statements into a statement about sets, and illustrate by a Venn diagram.

1   All dogs have four legs (D, L).

2   Some prime numbers are even (P, E).

3   The square of every negative number is positive (S, P).

4   Some polygons are regular polygons (P, R).

5   Every rational number is a real number (Q, R).

6   No prime number greater than 2 is even (P, E).

7   A scalar product is a real number (S, R).

In questions 8 to 12, write an implication equivalent to the given quantified statement (see Worked Example 1).

8   Any integer with zero in the units place is divisible by 5.

9   No surd has a decimal equivalent.

10   The base angles of an isosceles triangle are equal.

11a   Every even number greater than 2 can be expressed as the sum of two prime numbers. (Goldbach's conjecture which is still unproved.)

   b   State the converse and prove that it is not true by giving *one* counter-example.

12   Any year divisible by 4 is a leap year.
        Given that century years are leap years only if they are divisible by 400, write down a counter-example which disproves the above statement.

---

6   *Negation of quantified statements*

From a given statement we can form another statement which *denies* the given statement. For example, consider the statement

   257 *is a prime number.*

We can assert that this statement is false by writing
            (i) 257 is not a prime number,
    or   (ii) It is false that 257 is a prime number.
Each of the statements (i) and (ii) is called the *negation* of the given one.

Given that p represents a statement, the *negation* of p is written $\sim p$ (read as not-p), and is such that if p is true, $\sim p$ is false; if p is false, $\sim p$ is true.

To negate quantified statements, we must be careful about the meanings of the words 'some' and 'all'. Notice that there are just these two quantifiers, despite the various guises in which they appear. 'All' is called the *universal quantifier*, 'some' the *existential quantifier*. The last name comes from yet another way of expressing statements such as that of Worked Example 2 of Section 5, namely, 'There *exist* pilots who are women'.

Example 1. Negate the statement '*All cats have tails*'.

$$p: \quad All \text{ cats have tails.}$$

Since the given statement is taken to be true for *every* cat, the negation must assert that *at least one* cat has *no* tail, and so

$$\sim p: \quad Some \text{ cats do not have tails.}$$

Example 2. State the negation of '*Some pilots are women*'.

$$p: \quad Some \text{ pilots are women.}$$

Since the given statement asserts that there is *at least one* pilot who is a woman, the negation must assert that *all* pilots are *not*-women and so

$$\sim p: \quad \text{No pilots are women.}$$

An equivalent 'If ... then ...' negation is 'If x is a pilot, then x is not a woman'.

Example 3. Negate '*No pupils like examinations*'.
This can be written '*All* pupils dislike examinations'. Then as in Example 1, the negation is '*Some* pupils like examinations'.
From these examples, the following summary can be constructed.

Statement	Negation
(i) All As are Bs.	Some As are not-Bs.
(ii) Some As are Bs.	All As are not-Bs, or No As are Bs.

Exercise 6

1   For each of the Worked Examples 1, 2, 3 above construct an equivalent statement about sets, and illustrate this statement and its negation by Venn diagrams.

2   Which of the following is a negation of 'All boys are adventurous'?

a   No boys are adventurous.
b   All boys are unadventurous.
c   Some boys are not adventurous.
d   No boys are unadventurous.

3   Which of the following is a negation of 'No visitor may walk on the grass'?

a   All visitors may walk on the grass.
b   Some visitors may not walk on the grass.
c   All visitors may not walk on the grass.
d   Some visitors may walk on the grass.

Give a negation for each of the statements in questions *4* to *7*.

4   For all real x, x^2 is positive.

5   Some pupils find mathematics difficult.

6   No dogs like cats.

7   There exists a positive integer x such that $x + 3 > 0$.

Form a negation of each of the following statements, and say whether in your opinion it is the statement or the negation which is true.

8   Every parallelogram has half turn symmetry.

9   No schoolboy lies.

10   A number which has zero in the units place is divisible by five.

11   All numbers of the form $2^n - 1$, n an integer, are prime.

12   Some matrices have no multiplicative inverse.

13   For all real numbers x, $x^2 + 1$ is positive.

14   For all pairs of points A, B and their images A_1, B_1 under a translation, $\overrightarrow{AB} = \overrightarrow{A_1B_1}$.

15   For all lines DE joining the midpoints D of AB and E of AC in triangles ABC, DE is parallel to BC.

---

7   *Deductive reasoning*

In studying mathematics, we have discovered and used many mathematical truths called *theorems*. For example,

(i) 'The sum of the interior angles of a triangle is $180°$.'
(ii) '$x - h$ is a factor of polynomial $f(x) \Leftrightarrow h$ is a root of $f(x) = 0$.'
(iii) 'For all $x \in R$, $\sin^2 x + \cos^2 x = 1$.'

To prove a theorem, or a new result, its truth must be shown to follow from a set of other statements, each of which is accepted or has previously been proved true. Statements accepted as true without the need of justification are called *axioms*. For example, it is axiomatic that two different straight lines cannot meet in more than one common point.

In proving a theorem, or in deducing a result from known truths, use is made of valid patterns of argument based on logical principles of which the following are fundamental.

(1) *Principle of inference.* If p is true and $p \Rightarrow q$ is true, then q is true.

Schematically, we have:

If p, then q.	*or*	$p \Rightarrow q$
p is true.		p
Hence q is true.		Hence q

Example 1. If $f(x) = f(-x)$ for all $x \in R$, then $f : R \to R$ is an even function.
   $\cos x = \cos(-x)$ for all $x \in R$.
   Therefore the cosine is an even function.

(2) *Principle of the syllogism.* If $p \Rightarrow q$ is true and $q \Rightarrow r$ is true, then $p \Rightarrow r$ is true.

Schematically, we have:

$$\begin{array}{ccc} \text{If } p, \text{ then } q. & \quad or \quad & p \Rightarrow q \\ \text{and if } q, \text{ then } r. & & \text{and } q \Rightarrow r \\ \hline \text{Therefore if } p, \text{ then } r. & & \text{Therefore } p \Rightarrow r \end{array}$$

Example 2. If $x \in R$ and $x^2 - 4 = 0$, then $(x-2)(x+2) = 0$.
  If $(x-2)(x+2) = 0$, then $x = 2$ or $x = -2$.
Therefore if $x \in R$ and $x^2 - 4 = 0$, $x = 2$ or $x = -2$.

In both principles, the first two statements given as true are called *premises*, and the statement inferred from the premises is called the *conclusion*.

Patterns of argument which have a form which cannot be justified as valid are said to be invalid.

Example 3. The thieves were travelling in a dark-blue Aston Martin.
   The Smith brothers were travelling in a dark-blue Aston Martin.
Can a conclusion be deduced from these premises?

Re-writing each premise in 'If . . ., then . . .' form:
  If x is a thief, then x was travelling in a dark-blue Aston Martin.
  If x is a Smith brother, then x was travelling in a dark-blue Aston Martin.

The pattern is $p \Rightarrow q, r \Rightarrow q$ from which no deduction can be made. This fact can be verified by a Venn diagram.
  Let $T = \{\text{thieves}\}$, $S = \{\text{Smith brothers}\}$, $A = \{\text{travellers in dark-blue Aston Martins}\}$.
  Then $T \subset A$ and $S \subset A$, but there is no *necessary* connection between T and S. A possible Venn diagram is shown in Figure 5.

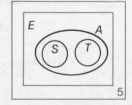

Additional examples of invalid arguments will be found in Exercise 7.

*   *   *   *

In mathematics, 'chains' of syllogisms are often encountered in proofs.

A number of mathematical relationships behave in the same way as implication; for example,

$$p = q \text{ and } q = r \quad \Rightarrow \quad p = r.$$

$$p > q \text{ and } q > r \quad \Rightarrow \quad p > r.$$

$$A \subset B \text{ and } B \subset C \quad \Rightarrow \quad A \subset C.$$

For the sake of brevity some proofs are expressed as chains of equalities (see Worked Example 1 in Section 8) or chains of implications (see Worked Example 2 in Section 8).

Indirect Proof. If we wish to prove that $p \Rightarrow q$ is true, we sometimes proceed by showing that $\sim q$ must be false and hence q must be true (see Worked Example 5 in Section 8).

Exercise 7

Examine the deductions in questions *1*–*8* and determine in each case whether or not the argument is valid.

1   If a quadrilateral is a parallelogram, its diagonals bisect each other. ABCD is a parallelogram.
Therefore the diagonals AC and BD bisect each other.

2   If $f(x) = x^n$, then $f'(x) = nx^{n-1}$.
$f(x) = x^3$.
Hence $f'(x) = 3x^2$.

3   If $f(x) = x^n$, then $f'(x) = nx^{n-1}$
$f'(x) = 3x^2$.
Hence $f(x) = x^3$.

4   If the equation of a curve is $y = ax^2 + bx + c$, $a \neq 0$, then it has one turning point.
The equation of a curve is $y = x^2 - 4x + 3$.
Therefore this curve has one turning point.

5   If the equation of a curve is $y = ax^2 + bx + c$, $a \neq 0$, then it has one turning point.
This curve has one turning point.
Therefore its equation is of the form $y = ax^2 + bx + c$.

6   If n is a prime number greater than 3, $(n+1)(n-1)$ is divisible by 24.
59 is a prime number greater than 3.
Therefore 3480 is divisible by 24.

7   If an angle is acute, then its cosine is positive.
An angle is acute if it is the smallest angle of a triangle.
Therefore the cosine of the smallest angle of a triangle is positive.

8   If two angles are right angles, the angles are equal.
Angle P = angle Q.
Therefore angles P and Q are right angles.

In questions *9–12*, draw a conclusion from the given premises if you can, and say whether the argument is valid.

9 *a* If a whole number is divisible by 6, then it is divisible by 3.
54 is divisible by 6.
Therefore ......

   b If a whole number is divisible by 6, then it is divisible by 3.
42 is divisible by 3.
Therefore ......

10   If mathematics is a useful subject, then studying mathematics is important.
If studying mathematics is important, then people should study mathematics.
Therefore ......

11   If x is a real number such that $x^2 - 3x + 2 = 0$, then $(x-1)(x-2) = 0$.
If $(x-1)(x-2) = 0$, then $x = 1$ or $x = 2$.
Therefore if x is a real number such that $x^2 - 3x + 2 = 0$, ...

12   If a quadrilateral is a square, it is a regular polygon.
All regular polygons can be inscribed in circles.
Therefore ......

13   p and q represent open sentences with the same set of replacements E, and with solution sets P and Q respectively. Find with the aid of a Venn diagram which of the following is a valid form of argument.

   a $p \Rightarrow q$ *b* $p \Rightarrow q$ *c* $p \Rightarrow q$
   Therefore $\sim q \Rightarrow \sim p$ Therefore $q \Rightarrow p$ Therefore $\sim p \Rightarrow \sim q$

14   Draw a Venn diagram and use it to verify that your answer to question *13* is correct when p is replaced by 'ABCD is a rectangle' and q by 'ABCD is a parallelogram'.

15   Investigate the validity of each of the following arguments.

a   If an angle is acute, its supplement is obtuse.
The supplement of angle A is not obtuse.
Hence angle A is not acute.

b   If an angle is acute, its supplement is obtuse.
Angle A is not acute.
Hence the supplement of angle A is not obtuse.

---

8   *Proof in mathematics*

In this course we have *proved* many results, some of which are shown below.

Algebra: The sum of the first n terms of a geometric series is given by
$$S_n = \frac{a(1-r^n)}{1-r}, \; r \neq 1.$$

Geometry: The point dividing the line joining $A(x_1, y_1)$ and $B(x_2, y_2)$ in the ratio $m:n$ is $\left(\dfrac{mx_2 + nx_1}{m+n}, \dfrac{my_2 + ny_1}{m+n} \right)$.

Trigonometry: In triangle ABC, $\dfrac{a}{\sin A} = \dfrac{b}{\sin B} = \dfrac{c}{\sin C}$.

Calculus: The derivative of $\sin x$ is $\cos x$.

As already stated in Section 7, to prove a new result, its truth must be shown to follow from other statements which are accepted as true (definitions, axioms and assumptions) and from theorems that have already been proved.

Five examples of mathematical proof follow.

Example 1. If $a, b \in R$, prove that $(a+b)^2 = a^2 + 2ab + b^2$.

Proof. $\begin{aligned}(a+b)^2 &= (a+b)(a+b) \quad \ldots \ldots \text{(definition of a power)}\\ &= (a+b)a + (a+b)b \ldots \ldots \text{(distributive law)}\\ &= (a^2 + ba) + (ab + b^2) \ldots \ldots \text{(distributive law)}\\ &= a^2 + (ba + ba) + b^2 \ldots \ldots \text{(associative law)}\\ &= a^2 + (ab + ab) + b^2 \ldots \ldots \text{(commutative law)}\\ &= a^2 + 2ab + b^2 \ldots \ldots \text{(definition of } 2 \times ab)\end{aligned}$

Example 2. If, in $\triangle ABC$, $\angle BAC$ is a right angle prove that $a^2 = b^2 + c^2$ (Figure 6).

Proof. In $\triangle$s ABC and ABD, $\angle B$ is common and $\angle BAC = 90° = \angle BDA$

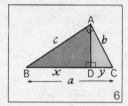

6

$\Rightarrow$ $\triangle$s ABC and ABD are equiangular

$\Rightarrow$ $\triangle$s ABC and ABD have corresponding sides in proportion.

$\Rightarrow$ $\dfrac{a}{c} = \dfrac{c}{x}$

$\Rightarrow$ $c^2 = ax.$

Similarly it can be proved in $\triangle$s ABC and ACD that $b^2 = ay.$

Hence $b^2 + c^2 = ay + ax = a(y + x) = a \cdot a = a^2.$

Example 3. Show that the point $(ap^2, 2ap)$, for each real number p, lies on the curve $y^2 = 4ax$.

Proof. $y^2 = (2ap)^2 = 4a^2p^2 = 4a \cdot ap^2 = 4ax$

Hence for each real number p, $(ap^2, 2ap)$ lies on the curve $y^2 = 4ax$.

Example 4. Prove that $\int \cos^2 x \, dx = \frac{1}{2}x + \frac{1}{4}\sin 2x + C.$

Proof. $\cos 2x = 2\cos^2 x - 1 \iff \cos^2 x = \frac{1}{2}(1 + \cos 2x)$
(already proved in trigonometry)
Hence $\int \cos^2 x \, dx = \frac{1}{2}\int (1 + \cos 2x)\, dx = \frac{1}{2}x + \frac{1}{4}\sin 2x + C.$
(by results proved in calculus)

Example 5. Prove that $\sqrt{2}$ is irrational.

Proof. Suppose that $\sqrt{2} = \dfrac{m}{n}$, where m and n are integers with no common factor. Square both sides.

$2 = \dfrac{m^2}{n^2} \Rightarrow m^2 = 2n^2 \Rightarrow m^2$ is even $\Rightarrow m$ is even $= 2k$ say $(k \in Z)$

$\Rightarrow$ $2n^2 = 4k^2 \Rightarrow n^2 = 2k^2 \Rightarrow n^2$ is even $\Rightarrow n$ is even.

$\Rightarrow$ m and n have a common factor 2.

But m and n have no common factor.

Therefore $\sqrt{2}$ cannot be expressed in the form $\dfrac{m}{n}$, and must be irrational.

This is an example of an *indirect proof*, where we demonstrate the truth of the statement p ($\sqrt{2}$ is irrational) by proving that $\sim p$ is false.

Exercise 8

1   Prove that the sum of the angles of $\triangle ABC$ is $180°$.
Draw MAN parallel to BC (Figure 7). Complete the following.

Proof. $p + x + q = \ldots$ (.................)
But $p = \ldots$ (.................)
and $q = \ldots$ (.................)
Hence $x + y + z = \ldots$

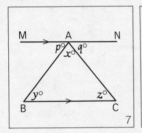

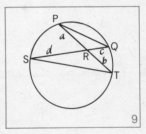

2   Prove that two angles $x°$ and $y°$ in the same segment of a circle are equal (Figure 8). (Use the angle $z°$ at the centre.)

3   Prove that, for two intersecting chords of a circle, $a \cdot b = c \cdot d$ (Figure 9 where $a = PR$, $b = RT$, $c = QR$, $d = RS$). (Use similar triangles PRQ and SRT.)

4   For three intersecting sets A, B and C, prove that $(A \cap B) \cup (A \cap C) = A \cap (B \cup C)$. (Draw a Venn diagram and use appropriate shading to find the regions represented by each side.)

5   Show that every point in the plane with coordinates of the form $x = a \cos 2\theta$, $y = a \sin 2\theta$, where $\theta \in R$ and a is a real constant, lies on the circle $x^2 + y^2 = a^2$.

6   Let M, R and Y represent the transformations of reflection in $y = x$, rotation of $\frac{1}{2}\pi$ radians about O and reflection in the y-axis respectively. Use matrix multiplication to show that $R \circ M = Y$.

7   Prove that $\displaystyle\int_a^b x^n\,dx + \int_b^c x^n\,dx = \int_a^c x^n\,dx$, $n \neq -1$. (Evaluate the integrals in terms of a, b and c.)

8   Prove that $\sqrt{3}$ is irrational. (Use an indirect proof; see Worked Example 5.)

9   If $y = mx + c$ is a tangent to the circle $x^2 + y^2 = r^2$, prove that $c = \pm r\sqrt{(1 + m^2)}$.

10   $P(x, y)$ is a point on the circle with a diameter joining the points $Q(x_1, y_1)$ and $R(x_2, y_2)$. Prove by means of gradients of perpendicular lines that the equation of the circle is
$$(x - x_1)(x - x_2) + (y - y_1)(y - y_2) = 0.$$

11   Prove the result in question *10* by means of the theorem of Pythagoras.

12   A diameter CD of a circle is perpendicular to a chord AB and cuts it at X.

Prove that AX = XB by completing the following:

$$
\begin{aligned}
\text{Under reflection in CD, the line of XA} &\leftrightarrow \dots\dots\dots\dots \\
\text{the semicircle CAD} &\leftrightarrow \dots\dots\dots\dots \\
\text{Therefore the intersection A} &\leftrightarrow \dots \\
\text{Also X} &\leftrightarrow \dots \\
\text{Therefore AX} &= \dots
\end{aligned}
$$

13   Two circles with centres X and Y intersect in A and B. Prove that XY bisects AB at right angles. (Use the common axis of symmetry of the two circles.)

14a   Prove that the square of every odd number is odd. (Take an odd number in the form $2m + 1$, and show that the square is of the form $2k + 1$.)

b   Prove also that if n^2 is odd, then n is odd ($n \in N$); use an indirect proof.

15   Show by replacing x by α and β, that α and β are roots of the equation $a(x - \alpha)(x - \beta) = 0$, $a \neq 0$. Show also that there cannot be a third root γ, different from both α and β. (Consider the factors of the product when x is replaced by γ.)

Summary

1   An *open sentence* is a sentence like $x - 5 = -1$ in which there is a variable x for which no particular replacement is specified.

A *statement* is a sentence conveying meaningful information, and is either true or false, e.g. 6 is greater than 5.

An *open sentence* becomes a *statement* for each possible replacement for its variable.

2   An *implication* is a sentence of the form '*If p, then q*',
e.g. if $2x = 6$, then $x = 3$.

3   The *converse of the implication* in 2 is '*If q, then p*',
i.e. if $x = 3$, then $2x = 6$.

4   A *counter-example* is an example which disproves the truth of a statement, e.g. if $ab = 0$, then $a = 0$.
Counter-example: $a = 1$, $b = 0$.

5   *Equivalence.* If $p \Rightarrow q$ and $q \Rightarrow p$ are both true, then $p \Leftrightarrow q$, i.e.
p is equivalent to q,
e.g. $\triangle ABC$ is right-angled at A $\Leftrightarrow a^2 = b^2 + c^2$.

6   *Quantified statements* contain the words *all, some, no*, e.g.

(i) *All* rectangles are parallelograms.

(ii) *Some* pilots are women.

(iii) *No* straight lines have equations $y = ax^2 + bx + c$, $a \neq 0$.

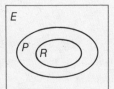

$P \cap R = R$

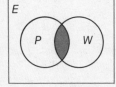

$P \cap W \neq \phi$

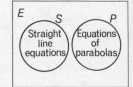

$S \cap P = \phi$

7 *Negation of a statement.* If statement p is true, its negation $\sim p$ is false, and vice versa.

Statement	*Negation*
All As are Bs.	Some As are not-Bs.
Some As are Bs.	No As are Bs.

8 *Deductions and proofs* can be made on the basis of statements which are known to be true, using valid patterns of argument based on logical principles,

e.g.

(i) $$\frac{\begin{array}{l} p \Rightarrow q \\ p \end{array}}{\text{Hence } q}$$

(ii) $$\frac{\begin{array}{l} p \Rightarrow q \\ \text{and } q \Rightarrow r \end{array}}{\text{Therefore } p \Rightarrow r}$$

Revision Exercises

Revision Exercise on Chapter 1
The Equations of a Circle

Revision Exercise 1

1 Find the equations of the circles with centre O, and:

a radius 12
b radius k
c radius half as long as that of the circle $x^2 + y^2 = 400$
d passing through the point $(1, -4)$.

2 Find the points of intersection of the circle centre O, radius 5, and the line through O with gradient 2.

3 A is the point $(8, 0)$ and B is $(2, 0)$. Given $\{P(x, y) : PA = 2PB\}$, find the equation of the locus of P, and describe it geometrically.

4 Describe in words the region given by $\{(x, y) : (x-1)^2 + (y+1)^2 < 9\}$.

5 Find the equations of the circles with centre $(2, 3)$, and:

a radius 6
b passing through the origin
c passing through the point $(3, -2)$
d touching the x-axis
e touching the y-axis.

6 Find the centre and radius, and hence the equation, of the circle passing through the vertices $O(0, 0)$, $A(0, 2)$, $B(6, 2)$, $C(6, 0)$ of the rectangle OABC.

7 Give the coordinates of the centres and the lengths of the radii of the circles:

a $x^2 + y^2 - 6x + 10y - 2 = 0$ b $3x^2 + 3y^2 + 6x - 12y + 1 = 0$

8   Find the equation of the image of the circle $x^2+y^2+4x+8y-5=0$ under:

   a reflection in the *x*-axis   *b* reflection in the origin

   c the translation $\begin{pmatrix} 2 \\ 4 \end{pmatrix}$.

9   By finding their centres and radii, show that the circles $x^2+y^2=4$ and $x^2+y^2-8x-6y+16=0$ touch each other externally.

10   Find the equation of the circle passing through the points $(3,0)$, $(-5,0)$ and $(0,-4)$.

11   Find the points where the line $y=5x+10$ cuts the circle $x^2+y^2-2x-4y-164=0$. Find also the equation of the diameter perpendicular to the given line.

12   Show that the point $A(-4,1)$ lies on the circle $x^2+y^2+10x+4y+19=0$, and find the equation of the tangent to the circle at A.

13   Find the points of intersection of the line $y=3$ and the circle $x^2+y^2=25$. Find the equations of the tangents to the circle at these points, and find the point of intersection of the tangents.

14*a*   Verify that $C(8,7)$ lies on the circle $x^2+y^2-10x-7y+16=0$.
   b   Calculate the length of the chord AB cut off by the *x*-axis.
   c   If A is nearer the origin than B, show that AC is a diameter of the circle.

15*a*   Find the centres and radii of the circles $x^2+y^2-14x+6y-42=0$ and $x^2+y^2+10x-12y+36=0$, and show that the circles touch externally.
   b   Use the ratio of radii to calculate the coordinates of the point of contact of the circles, and find the equation of the common tangent at that point.

16   Find the equations of the tangents to the circle $x^2+y^2=36$ from the point $(8,0)$.

17   Find the equations of the tangents to the circle $x^2+y^2-8x-4y+16=0$ from the point $(0,2)$.

18   Show that the line $x+9=0$ is a tangent to the circles $x^2+y^2=81$ and $x^2+y^2+10x+24y+153=0$, and find the points of contact.

19   Show that the tangents to the circle $x^2 + y^2 - 4x - 2y - 20 = 0$ at A(7, 1) and B(-1, 5) intersect at the point (7, 11).

20   Show that the tangent to the circle $x^2 + y^2 = 10$ at the point A(-3, -1) is also a tangent to the circle $x^2 + y^2 - 8x + 4y - 20 = 0$, and find the point of contact B. Calculate the length of the common tangent AB.

21   A is the point (2, 8), B is (-4, 2) and C is (-4, 8). Show that angle ACB is a right angle, and find the equation of the circumcircle of △ABC. Prove that this circle touches the line $y = x$, and find the point of contact.

22   Prove that the line $x \cos \alpha + y \sin \alpha = 3$ is a tangent to the circle $x^2 + y^2 = 9$ for all values of α. What is the point of contact?

23   Find the equations of the tangents to the circle $x^2 + y^2 - 4x - 2y + 4 = 0$ from the origin.

24   Find the equation of the tangent at (2, 1) to the circle $x^2 + y^2 = 5$. If this tangent is also a tangent to the circle $(x - 3)^2 + (y - 2)^2 = a^2$, find the value of a^2.

25   Given constants a, where $a > 0$, and θ, where $0 < \theta < \frac{1}{2}\pi$, find the centre and radius of $x^2 + y^2 - 2ax \sin \theta - 2ay \cos \theta + a^2 \cos^2 \theta = 0$ in terms of a and θ. Show that the circle touches the y-axis, and from the symmetry of the figure, or otherwise, find the equation of the other tangent through the origin to the circle. What is the equation of the tangent when $\theta = \frac{1}{4}\pi$?

Revision Exercise on Chapter 2 *Composition of Transformations 2*

Revision Exercise 2

1   △ABC has vertices A(2, 0), B(4, 0), C(4, 1). Use matrices to find the coordinates of the images of A, B and C under:

   a reflection in the x-axis   *b* a rotation about O of π radians.

2   A square has vertices O(0, 0), A(1, 0), B(1, 1), C(0, 1). Find the coordinates of the images of O, A, B and C under the transformation with matrix $\begin{pmatrix} 2 & -1 \\ 1 & 4 \end{pmatrix}$. Show the square and its image in a diagram.

3   Verify that the points $(1, 1)$ and $(2, 3)$ lie on the line $y = 2x - 1$, and find their images under the transformation with matrix $\begin{pmatrix} 3 & 1 \\ -1 & 1 \end{pmatrix}$. Assuming that the image of the given line is also a straight line, find its equation.

4   Quadrilateral ABCD has vertices $A(0, 2), B(2, 0), C(0, -2), D(-2, 0)$. Find the images of A, B, C and D under a clockwise rotation about O of $\frac{3}{2}\pi$ radians. Explain the result geometrically.

5   Rhombus OABC has vertices $O(0, 0), A(6, 2), B(8, 8), C(2, 6)$. Find the images O_1, A_1, B_1, C_1 of the vertices under the dilatation $[O, 2]$. What dilatation and what matrix applied to OABC would produce a rhombus one quarter the size of OABC?

6   A and B are matrices associated with reflection in the line $y = x$ and a clockwise rotation about O of $\frac{1}{2}\pi$ radians respectively. Find the matrix products AB and BA.

If $X = \begin{pmatrix} a & b \\ c & d \end{pmatrix}$ and $XAB = BA$, find a, b, c and d.

7   A composite mapping determined by the equations

$\begin{aligned} x_1 &= x - 2 \\ y_1 &= y + 2 \end{aligned}$ and $\begin{aligned} x_2 &= 4x_1 + 8 \\ y_2 &= 4y_1 - 8 \end{aligned}$ maps $A(x, y)$ to $B(x_2, y_2)$.

Find the single geometrical transformation and associated matrix that effect this composite mapping.

8   $A = \begin{pmatrix} 1 & 0 \\ 0 & -1 \end{pmatrix}$, $B = \begin{pmatrix} 0 & -1 \\ -1 & 0 \end{pmatrix}$ and $C = \begin{pmatrix} 0 & -1 \\ 1 & 0 \end{pmatrix}$. Calculate:

a ABC b BAC c CBA.

Describe each result geometrically.

9   The transformation of reflection in a certain line through the origin is associated with the matrix $\begin{pmatrix} \cos 2\theta & \sin 2\theta \\ \sin 2\theta & -\cos 2\theta \end{pmatrix}$.

Show that under this transformation the point $(a \cos \theta, a \sin \theta)$ maps onto itself. Deduce that the equation of the axis of reflection is $y = x \tan \theta$, and state the angle it makes with OX.

10a What transformation is associated with $\begin{pmatrix} \cos\alpha & -\sin\alpha \\ \sin\alpha & \cos\alpha \end{pmatrix}$?

  b Show by matrix multiplication that

$$\begin{pmatrix} \cos\alpha & -\sin\alpha \\ \sin\alpha & \cos\alpha \end{pmatrix}^2 = \begin{pmatrix} \cos 2\alpha & -\sin 2\alpha \\ \sin 2\alpha & \cos 2\alpha \end{pmatrix}.$$

Explain this result geometrically, and write down the corresponding matrix for $\begin{pmatrix} \cos\alpha & -\sin\alpha \\ \sin\alpha & \cos\alpha \end{pmatrix}^n$, where n is a positive integer.

11 $\triangle$ABC has vertices A(2, 1), B(4, 1), C(4, 2). Show in a diagram the triangle and its images under:

  a the reflection in line $y = x$, followed by the translation $\begin{pmatrix} 6 \\ 0 \end{pmatrix}$

  b the translation $\begin{pmatrix} 6 \\ 0 \end{pmatrix}$, followed by reflection in $y = x$.

Is the order important?

12 $\triangle$PQR has vertices P(4, 2), Q(4, 0), R(3, 0). Show in a diagram the triangle and its images under:

  a a clockwise rotation about O of $\frac{1}{2}\pi$ radians, followed by reflection in the line $y = x$

  b reflection in $y = x$ followed by a clockwise rotation about O of $\frac{1}{2}\pi$ radians.

   Is the order important? State the single transformation which is equivalent to each of the composite transformations.

13 Matrix Q represents a positive quarter turn about O, and matrix M_2 represents reflection in the line $y = x$. If $M_2 M_1 = Q$, show by using an inverse matrix that M_1 must represent reflection in the x-axis.

14 Q, Y and M are matrices associated with an anticlockwise quarter turn about O, reflection in the y-axis and reflection in the line $y = x$ respectively.

  a Show that $Q = YM$ and that $Q^{-1} = M^{-1}Y^{-1}$.

  b Express $Y^{-1}M^{-1}$ as a single matrix and interpret it geometrically.

15a Given $M = \begin{pmatrix} \cos 2\theta & \sin 2\theta \\ \sin 2\theta & -\cos 2\theta \end{pmatrix}$, find M^{-1}, M^2 and $M^3 (= M \cdot M^2)$.

b Deduce matrix M^n if: (1) n is even (2) n is odd.

16 Show that if A, B, X and H are 2×2 matrices which have inverses, and $ABX = H$, then $X = B^{-1}A^{-1}H$. (Begin by pre-multiplying each side by A^{-1}.)

If A, H and B represent reflection in the y-axis, a half turn about O, and reflection in $y = x$ respectively, calculate $B^{-1}A^{-1}H$, and hence interpret X geometrically.

17 A rectangle has vertices $O(0, 0)$, $A(2, 0)$, $B(2, 1)$, $C(0, 1)$. Show in a diagram the rectangle and its images under:

a a dilatation $[O, 3]$, followed by a half turn about O
b the half turn followed by the dilatation.

Find the matrices D and H associated with these transformations, and verify that $HD = DH$. Show also that $(HD)^{-1} = D^{-1}H^{-1}$.

18 A square has vertices $O(0, 0)$, $A(1, 0)$, $B(1, 1)$, $C(0, 1)$. Find the image of the square under the composite transformation of an anti-clockwise rotation about O of $\frac{1}{4}\pi$ radians followed by a dilatation with centre the origin and scale factor $\sqrt{2}$. Illustrate in a diagram. If D and R are the corresponding matrices, show that $(DR)^{-1} = R^{-1}D^{-1}$.

19 Find the equation of the image of the line $2x - y = 7$ under the transformation with matrix $\begin{pmatrix} 2 & -1 \\ 0 & 1 \end{pmatrix}$.

20 Find the equation of the image of the line $2x + y = 4$ under the transformation with matrix $\begin{pmatrix} 1 & -1 \\ 1 & 1 \end{pmatrix}$. Verify that $A(2, 0)$ lies on $2x + y = 4$ and find its image A_1. Show that O is an invariant point under the transformation, and state the size of $\angle AOA_1$.

21 Find the equation of the image of the circle $x^2 + y^2 = 9$ under the transformation given by $x_1 = -x + y$, $y_1 = 2x - y$.

22a Find the equation of the image of the circle, centre O, radius 5, under the transformation with matrix $\begin{pmatrix} 2 & 0 \\ 0 & 1 \end{pmatrix}$.

b Find the images of $(5, 0)$, $(-5, 0)$, $(0, 5)$, $(0, -5)$ under the transformation, and hence sketch the circle and its image on the same diagram.

23*a* Show that the matrix $M = \begin{pmatrix} 1 & 0 \\ k & 0 \end{pmatrix}$ maps all points on the x, y plane onto the line $y = kx$, where k is a constant.

b Find M^2 and M^3, and explain why M^{-1} does not exist.

24 A square has vertices $O(0, 0)$, $A(1, 0)$, $B(1, 1)$, $C(0, 1)$. Find the images O, A_1, B_1, C_1, of the vertices under the transformation with matrix $M = \begin{pmatrix} a & b \\ c & d \end{pmatrix}$. Show by gradients that $OA_1B_1C_1$ is a parallelogram.

By dividing up the rectangle which has diagonal OB_1 and two adjacent sides on OX and OY, show that the magnitude of the area of $OA_1B_1C_1$ is equal to the determinant of the matrix M.

25 The matrix $\begin{pmatrix} p & q \\ r & s \end{pmatrix}$ maps (x, y) to (x_1, y_1). Prove that the image of the line $ax + by + c = 0$ is a straight line.

26 $\triangle ABC$ has vertices $A(0, 3)$, $B(0, 4)$, $C(2, 4)$. Show in a diagram the triangle and its image $A_1B_1C_1$ under a clockwise quarter turn about O followed by a translation $\begin{pmatrix} 0 \\ -6 \end{pmatrix}$.

Find the equations of the perpendicular bisectors of AA_1 and BB_1, and hence find the centre of the single rotation that maps $\triangle ABC$ to $\triangle A_1B_1C_1$.

27 The matrices M and R are associated with reflection in $y = -x$ and an anticlockwise rotation about O of $\frac{1}{2}\pi$ radians respectively.

a Show that $MR \neq RM$, and find matrix X such that $XMR = RM$.

b Find the equations of the images of the parabola $y = x^2 - 5x + 4$ under the transformations given by MR, RM and X. Illustrate by a diagram.

Revision Exercise on Chapter 3
Mathematical Deduction and Proof

Revision Exercise 3

1 Which of the following statements are true, and which are false, for all $a, b, c \in Q$?

a $a+b = a+c \Rightarrow b = c$ b $ab = ac \Rightarrow b = c$

c $a < b \Rightarrow a+c < b+c$ d $a < b \Rightarrow ac < bc$

    Amend those statements which are false to give true statements.

2 Copy and complete the following to give true implications:

a If $n \in Z$, $n(n+1) = n + 361 \Rightarrow n = \ldots$
b If $X = \{x : 1 < x \leqslant 4\}$ and $Y = \{x : 3 < x \leqslant 6\}$, then $X \cup Y = \ldots$
c If $x \in R$, $(x-3)(x+2) \leqslant 0 \Rightarrow \ldots$
d If a parallelogram has an axis of symmetry $\ldots$

3 Given $x \in R$, $x(x-3) = 0 \Rightarrow x = 0$. Is this a true implication?
    What is the converse? Is the converse true?

4 State the converse of each of the following:

a Every mapping is a relation.
b If a triangle has angles of $40°$ and $60°$, it has an angle of $80°$.
c All rational numbers are real numbers.
d If $x \in R$, then $x > 1 \Rightarrow x^2 > x$.

    Now show, by giving one counter-example in each case, that every one of the converses you have stated is false.

5 Prove by a counter-example that each of the following is false:

    If $x, y \in R$:   a $x > y \Rightarrow x^2 > y^2$ b $x < y \Leftrightarrow \dfrac{1}{x} > \dfrac{1}{y}$

    Amend a and b so that they are true.

6 Put the appropriate symbol, $\Rightarrow$ or $\Leftrightarrow$, in each of the following:

a $\triangle$ABC is congruent to $\triangle$DEF $\ldots$ $\triangle$ABC is similar to $\triangle$DEF.
b $\cos^2 a° + \sin^2 a° = 1 \ldots \cos^2 a° = 1 - \sin^2 a°$.
c m is odd and n is odd $\ldots$ mn is odd $(m, n \in Z)$.
d $f'(a) = 0 \ldots f(a)$ is a stationary value of f.

7   Which of the following is a negation of the statement:
    'All sine and cosine functions are differentiable'?
    A   No sine or cosine functions are differentiable.
    B   All sine and cosine functions are not differentiable.
    C   Some sine and cosine functions are not differentiable.
    D   Some sine and cosine functions are differentiable.

8   In a certain school there is a rugby fifteen, a cricket eleven, and a swimming eight. 3 boys are in all three teams, 9 boys are in only the rugby team, and 5 boys are in both the rugby and cricket teams; 2 boys are in the swimming team only. Deduce the number of boys who represent the school at cricket only.

9   Complete the following reasoning:
    If a set S has n elements, it has 2^n subsets.
    $A = \{1, 3, 5, 7, 11\}$ has ... elements.
    Therefore...........................

10   p, q, and r represent straight lines. Investigate the conclusion in each of the following arguments.

    a p is parallel to q. b p is perpendicular to q.
       q is parallel to r. q is perpendicular to r.
       Therefore p is parallel to r.   Therefore p is perpendicular to r.

11   In the following arguments, say whether or not the conclusion necessarily follows from the premises. Draw Venn diagrams in each case to check your answer.

    a Boys who waste their time at school fail their examination.
       Peter failed his examination.
       Therefore Peter wasted his time at school.

    b All rectangles are cyclic quadrilaterals.
       ABCD is a rectangle.
       Therefore ABCD is a cyclic quadrilateral.

12   Make up some valid arguments based on the following premises:

    a All plant tissues are made up of cells.
    b A buttercup is a plant.
    c All cells contain a living substance called protoplasm.

13   Use an indirect argument to show that if two straight lines a and b in the same plane are perpendicular to a third line c in the plane then a is parallel to b.

Cumulative Revision Section

Cumulative Revision Section (Books 8-9)

Revision Topic 1. The Gradient and Equations of a Straight Line

Reminders

1 a The gradient of the line joining the points (x_1, y_1) and (x_2, y_2) is $\dfrac{y_2 - y_1}{x_2 - x_1}$, provided that $x_2 \neq x_1$.

b A line parallel to the y-axis has no defined gradient.

c The gradient of a line is equal to $\tan \theta$, where θ is the angle which the line makes with the positive direction of the x-axis.

2 a The line through the point $(0, c)$ with gradient m has equation $y = mx + c$.

b The line through the point (a, b) with gradient m has equation $y - b = m(x - a)$.

c An equation of the form $Ax + By + C = 0$ (A and B not both zero) is the equation of a straight line.

d $x = h$ is the equation of a line parallel to the y-axis.

e $y = k$ is the equation of a line parallel to the x-axis.

3 If m_1 and m_2 are the gradients of two lines then:

a the lines are parallel $\Leftrightarrow m_1 = m_2$

b the lines are perpendicular (but not parallel to the axes) $\Leftrightarrow m_1 m_2 = -1$.

4 A is the point (x_1, y_1) and B is the point (x_2, y_2).

a The length of AB $= \sqrt{[(x_2 - x_1)^2 + (y_2 - y_1)^2]}$.

b If P divides AB in the ratio $m:n$, $x_P = \dfrac{mx_2 + nx_1}{m + n}$, $y_P = \dfrac{my_2 + ny_1}{m + n}$.

Exercise 1

1 Calculate where possible the gradients of the straight lines joining the following pairs of points.

    a $(-4, 1), (1, 3)$ *b* $(6, 2), (4, 4)$ *c* $(1, 5), (1, -2)$

    d $(-3, -4), (-1, -8)$ *e* $(3, 6), (-3, 6)$ *f* $(p, q), (h, k)$

2 Write down the equation of the straight line through the point $(0, -3)$ with gradient $-\frac{1}{2}$.

3 Find the equations of the straight lines joining the pairs of points in question *1*, *a*, *b* and *c*.

4 Two lines have equations $2x - 3y + 4 = 0$ and $4x - 6y = 7$. Express the equations in the form $y = mx + c$. Explain why the lines are parallel.

5 Explain why the lines with equations $3x - y - 4 = 0$ and $x + 3y + 1 = 0$ are perpendicular.

6 A is the point $(1, 2)$, B is $(-2, -2)$ and C is $(-6, 1)$. Show that $\angle ABC$ is right: *a*   by gradients   *b*   by the distance formula.

7 $A(7, 3)$, $B(8, 10)$ and $C(-4, 1)$ are vertices of a triangle. AD is an altitude of the triangle. Find: *a*   the equation of AD   *b*   the coordinates of D.

    Show also that the length of AD is 5.

8 The line through $A(0, 4)$ with gradient 2 meets the line through the points $(1, 0)$ and $(1, -1)$ at B. Calculate the coordinates of B.

9 The vertices of a triangle are $A(1, 10)$, $B(5, 2)$ and $C(9, 6)$.

    a   Find the equation of the altitude AD.

    b   Find the coordinates of the midpoint M of BC and the equation of the median AM.

    c   Explain why the altitude and the median have the same equation.

10*a* Find the equation of the straight line passing through the points $A(-4, 2)$ and $B(6, 1)$.

    b   Write down the images of A and B under a positive quarter turn about the origin.

    c   Hence, or otherwise, find the equation of the image of AB.

11a   Show that the midpoint of the line joining the points $P(b-2, a+2)$ and $Q(a, b)$ lies on the line with equation $y = x + 2$.

  b   Prove that PQ is perpendicular to the line $y = x + 2$.

  c   Hence, or otherwise, find the image of the point $(5, 2)$ under reflection in $y = x + 2$.

12a   Find the equation of the straight line through $(5, -2)$ perpendicular to $3x - 2y = 5$.

  b   The straight line you have found cuts the x- and y-axes at A and B respectively. Find the coordinates of A and B.

  c   Find the ratio in which the straight line $y = x$ cuts AB.

13   A is the point $(4, 6)$, B is $(-2, -2)$ and C is $(2, -4)$. If D is the midpoint of AB and E is the midpoint of AC, show that DE is parallel to BC and that $DE = \frac{1}{2}BC$.

14   Find the equation of the straight line joining the points $A(1, 4)$ and $B(5, 6)$. Hence or otherwise show that AB is bisected by the line $x + 3y - 18 = 0$.

15   P is the point $(0, 3)$ and Q is $(4, 4)$. Find the ratio in which PQ is cut by the line $x + 4y = 24$.

16a   A is the point $(3, -2)$ and B is $(7, 4)$. Use the distance formula to find the equation of the locus $\{P : PA = PB\}$.

  b   State why this locus must be a straight line.

  c   Find the equation of the straight line through the midpoint of AB and perpendicular to AB. What do you notice about the two equations?

17   OP is the perpendicular from the origin to a straight line AB and meets AB at P. If $OP = p$ and $\angle XOP = \alpha$ radians, express the coordinates of P in terms of p and α.

      Show that the equation of AB can be written in the form $x \cos \alpha + y \sin \alpha = p$.

18   The vertices of a triangle are $A(2, -1)$, $B(5, 5)$ and $C(-2, 7)$. P divides AB in the ratio $1:2$, Q divides AC in the ratio $3:1$, and R divides BC externally in the ratio $6:1$.

      Find the coordinates of P, Q and R, and show that these points are collinear.

Revision Topic 2   *Vectors 2*

Reminders

1 a   A vector in three dimensions can be represented by a directed line segment such as $\overrightarrow{AB}$ and can be defined by a number triple.

b   Every vector in space can be expressed uniquely in terms of three non-coplanar vectors which form the *basis* for the whole set of vectors.

If $r = lu + mv + nw$, l, m and n are the components of r, and

$$r = \begin{pmatrix} l \\ m \\ n \end{pmatrix}.$$

c   The basis used in connection with rectangular coordinates is i, j and k, vectors of unit length in the directions of the axes OX, OY and OZ respectively.

2 a   The position vector of a point C dividing AB in the ratio $m:n$ is given by $c = \dfrac{mb + na}{m + n}$.

b   In particular, the position vector of the midpoint of AB is $\frac{1}{2}(a + b)$.

c   If P divides the line joining A(x_1, y_1, z_1) and B(x_2, y_2, z_2) in the ratio $m:n$, then $x_P = \dfrac{mx_2 + nx_1}{m + n}$, $y_P = \dfrac{my_2 + ny_1}{m + n}$, $z_P = \dfrac{mz_2 + nz_1}{m + n}$ (the Section Formulae).

3 $|\overrightarrow{AB}| = \sqrt{[(x_2 - x_1)^2 + (y_2 - y_1)^2 + (z_2 - z_1)^2]}$ (the Distance Formula).

4 a   The scalar (dot) product of two vectors a and b is defined by $a.b = |a\|b|\cos\theta$ or $a_1 b_1 + a_2 b_2 + a_3 b_3$.

b $\cos\theta = \dfrac{a.b}{|a\|b|} = \dfrac{a_1 b_1 + a_2 b_2 + a_3 b_3}{\sqrt{(a_1{}^2 + a_2{}^2 + a_3{}^2)(b_1{}^2 + b_2{}^2 + b_3{}^2)}}.$

c $a.b = b.a$ and $a.(b + c) = a.b + a.c$ (the Distributive Law).

Exercise 2

1   A, B and C are the points $(2, -1, 3), (4, 6, -2)$ and $(0, 2, 4)$ respectively. Express in component form the vectors represented by:

a $\overrightarrow{OB}$ *b* $\overrightarrow{AB}$ *c* $\overrightarrow{BA}$ *d* $\overrightarrow{AB} + \overrightarrow{BC}$ *e* $\overrightarrow{OA} - \overrightarrow{OC}.$

2   The points $P(3, 4, 6)$, $Q(4, 2, 2)$ and $R(6, h, k)$ are collinear. Find h and k. In what ratio does Q divide PR?

3   If M is the point $(2, -1, -3)$ and N is $(1, 4, -2)$, express the vector represented by $\overrightarrow{MN}$ in terms of i, j and k.

4   A is the point $(-1, 3, 4)$ and B is $(2, 1, -2)$.

    a Calculate the distance between A and B.
    b If P divides AB in the ratio $2:1$, calculate the coordinates of P.

5 *a* Express the vectors $a = \begin{pmatrix} 1 \\ 4 \\ -2 \end{pmatrix}$, $b = \begin{pmatrix} 6 \\ -2 \\ -1 \end{pmatrix}$ and $c = \begin{pmatrix} 12 \\ -4 \\ -2 \end{pmatrix}$ in

    terms of i, j and k.

    b Show that a is perpendicular to b.
    c Show that b is parallel to c.

6   ABCDEF is a regular hexagon with centre O. $\overrightarrow{AB}$ and $\overrightarrow{BC}$ represent vectors u and v respectively. Express in terms of u and v the vectors represented by:

    a $\overrightarrow{AD}$ *b* $\overrightarrow{CD}$ *c* $\overrightarrow{DE}$ *d* $\overrightarrow{EF}$ *e* $\overrightarrow{FA}$ *f* $\overrightarrow{OA}$

    g $\overrightarrow{OB}$ *h* $\overrightarrow{OC}$ *i* $\overrightarrow{OD}$ *j* $\overrightarrow{OE}$ *k* $\overrightarrow{OF}$.

    If a side of the hexagon is 5 units long, evaluate $u \cdot v$.

7   P is the point $(3, 4, 1)$, Q is $(9, 1, -5)$ and R is $(11, 0, -7)$.

    a Prove that P, Q and R are collinear.
    b Find the ratio in which Q divides PR.
    c Find the ratio in which P divides RQ.

8   The position vectors of A, B, C, D are $\begin{pmatrix} 4 \\ -8 \\ 4 \end{pmatrix}$, $\begin{pmatrix} 1 \\ -2 \\ 1 \end{pmatrix}$, $\begin{pmatrix} -12 \\ 24 \\ -12 \end{pmatrix}$,

    $\begin{pmatrix} 1 \\ 4 \\ k \end{pmatrix}$ with respect to an origin O. What can you say about A, B

    and C?

    If the vector represented by $\overrightarrow{AB}$ is perpendicular to the vector represented by $\overrightarrow{OD}$, find k.

9   ABCDEFGH is a parallelepiped in which AB = 1 unit, AD = 2 units
     and DH = 3 units. $\overrightarrow{AB}$ represents a, $\overrightarrow{AD}$ represents b and $\overrightarrow{AE}$
     represents c. The angles between these vectors, taken in pairs, are
     each 60°:

a   Evaluate   (1) $a.b$ (2) $a.c$ (3) $b.c$ (4) $a.a$.

b   Use the distributive law to evaluate $(a+b+c).(a+b+c)$, and then
     write down the length of AG.

c   Evaluate $a.(a+b+c)$ and show that $\cos \angle BAG = 0.7$.

10   ABC is a triangle and squares BCDE and ABFG are drawn outside
     the triangle.

a   Prove that $\angle FBE$ is the supplement of $\angle ABC$.

b   Let $\overrightarrow{BF}$, $\overrightarrow{BA}$, $\overrightarrow{BC}$ and $\overrightarrow{BE}$ represent vectors p, q, r and s respectively.
     Use the scalar product and the distributive law to prove that FC is
     perpendicular to AE.

11   PQRS is a parallelogram. Express s, the position vector of S, in
     terms of p, q, r.
        M is the midpoint of PQ, and N is the midpoint of SR. Use position
     vectors to show that $\overrightarrow{PN} = \overrightarrow{MR}$. What kind of figure is PNRM?

12   Triangle PQR has vertices $P(3,5,0)$, $Q(1,3,-1)$ and $R(-1,4,1)$.
     Prove that the triangle is right-angled and isosceles, using:

a   the distance formula only   b   the scalar product only.

13   $u = \begin{pmatrix} a_1 \\ a_2 \\ a_3 \end{pmatrix}$, $v = \begin{pmatrix} 3 \\ -2 \\ -1 \end{pmatrix}$ and $w = \begin{pmatrix} 1 \\ 3 \\ -4 \end{pmatrix}$. If vector u has unit length,

     and is perpendicular to vectors v and w write down three equations
     satisfied by a_1, a_2 and a_3.
        Show that $a_2 = a_3$, and hence find possible values of a_1, a_2 and a_3.

14   Which of the following statements are true and which are false?

a $(a.b)c$ is a real number.
b $3a.b = 0 \Leftrightarrow a$ is perpendicular to b.
c $a.b = a.c \Leftrightarrow b = c$.
d   If a, b and c are three independent vectors in three dimensions then,
     given a fourth vector d, real numbers k, l and m can be found such
     that $d + ka + lb + mc = 0$.

Cumulative Revision Exercises

Exercise 1

1   Show the set $A = \{(x, y): x^2 + y^2 = 25\}$ on squared paper. Draw also the image of A under reflection in the line $x = -3$, and state the coordinates of the centre of the image.

2   In Figure 1, the equation of the circle is $x^2 + y^2 = 64$, and the line PQ cuts the circle at A and B on the axes as shown.

   a Find the equation of PQ.
   b. State in set notation the locus indicated by the shaded area.
   c Calculate, in terms of π, the area of the shaded part.

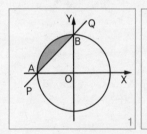
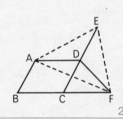
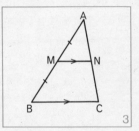

3   In Figure 2, ABCD is a parallelogram. BC = CF; CD = DE. If $\overrightarrow{BA}$ represents the vector $\boldsymbol{u}$ and $\overrightarrow{BC}$ represents the vector $\boldsymbol{v}$, express in terms of $\boldsymbol{u}$ and $\boldsymbol{v}$ the vectors represented by $\overrightarrow{AE}$, $\overrightarrow{DF}$, $\overrightarrow{EF}$ and $\overrightarrow{FA}$. Check that $\overrightarrow{AE} + \overrightarrow{EF} + \overrightarrow{FA} = 0$.

4   In Figure 3, M is the midpoint of AB, and MN∥BC. $\overrightarrow{AB}$ represents vector $\boldsymbol{c}$, and $\overrightarrow{AC}$ represents $\boldsymbol{b}$. Show that $\overrightarrow{MN}$ represents $\frac{1}{2}(\boldsymbol{b} - \boldsymbol{c})$. What can you say about N?

5 a P is a point outside a circle, centre O. PS and PT are tangents touching the circle at S and T. Prove that quadrilateral OSPT is cyclic.
   b If P is a point outside a circle, centre O, describe how you could construct two tangents from P to the circle.

Exercise 2

1   ABCD is a parallelogram in which $\overrightarrow{AB}$ represents u and $\overrightarrow{AD}$ represents v.   AC and BD intersect in H.   P is a point on AB such that AP:PB = 1:3 and PH produced meets DC in Q. Find in terms of u and v the vectors represented by $\overrightarrow{DQ}$ and $\overrightarrow{PH}$.

2   A and B are the points $(-1,1)$ and $(3,4)$ respectively.
  a Write down the components of the position vectors a and b.
  b Write down the components of the vector represented by $\overrightarrow{AB}$.
  c Calculate the magnitude of $b - a$.

3   A is the point $(4, -1)$, B is $(11, 5)$ and C is $(6, 2)$. Find the coordinates of D such that: a $\overrightarrow{CD} = 2\overrightarrow{AB}$ b $\overrightarrow{CD} = 2\overrightarrow{BA}$.

4 a Write down the equation of the line L joining the origin to the point A(4, 2). What is the gradient of this line?

  b Find: (1) A_1, the image of A under reflection in the x-axis
           (2) the equation of L_1, the image of L under reflection in the x-axis.

  c If L_2 is the image of L_1 under reflection in the line $y = 2$, give a transformation which maps L to L_2, and find the equation of L_2.

5   Find the 2×2 matrix which maps P(x, y) to P$_1(x + y, x - y)$. Find the image of the set of points on: a the x-axis b the y-axis.

6   Two tangents AB and AC are drawn to a circle centre O and radius 10 cm. AO = 25 cm. Use your tables to calculate $\angle$ BAC.

7   In Figure 4, $\angle$ AOB = 138°. Calculate the size of $\angle$ BDC.

8   Copy Figure 5; mark four pairs of equal angles at the circumference. Prove the opposite angles of a cyclic quadrilateral supplementary.

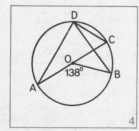
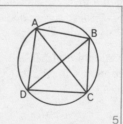
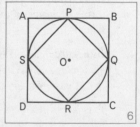

4   5   6

9 a Show that area of square ABCD: area of circle: area of square PQRS = 4:π:2 in Figure 6.

  b If $T \circ V$ maps PQRS to ABCD, and V is a rotation of $+45°$ about O, what transformation does T represent?

Exercise 3

1 *a* Find the equations of the straight lines through the point A$(-2, 3)$ which are parallel and perpendicular to the straight line $2x - 3y + 4 = 0$.

   b The two lines you have found cut the line $11x + 3y = 26$ at B and C respectively. Calculate the coordinates of B and C.

   c Write down the coordinates of M, the midpoint of BC, and the coordinates of D, the fourth vertex of rectangle ABDC.

2 *a* Find the centre and the radius of the circle with equation $x^2 + y^2 - 8x + 6y = 0$. State why the circle passes through O.

   b Find the coordinates of the other point P in which the circle cuts the *x*-axis.

   c Find the equations of the tangents to the circle at O and P.

   d Calculate the coordinates of Q, the point of intersection of these two tangents and verify by the distance formula that OQ = PQ.

3 *a* P divides the join of B$(5, -2, 1)$ and C$(-3, 22, -15)$ in the ratio $3:5$. Find the coordinates of P.

   b If A is the point $(-3, 6, -3)$, express the vectors represented by $\overrightarrow{AB}$, $\overrightarrow{AP}$ and $\overrightarrow{AC}$ in component form, and show that $|\overrightarrow{AB}| : |\overrightarrow{AC}| = |\overrightarrow{BP}| : |\overrightarrow{PC}|$.

   c Prove that $\angle\,BAP = \angle\,CAP$.

4 *a* Which geometrical transformations are associated with the matrices
$$I = \begin{pmatrix} 1 & 0 \\ 0 & 1 \end{pmatrix},\ P = \begin{pmatrix} 0 & -1 \\ 1 & 0 \end{pmatrix},\ Q = \begin{pmatrix} 0 & 1 \\ -1 & 0 \end{pmatrix} \text{ and } H = \begin{pmatrix} -1 & 0 \\ 0 & -1 \end{pmatrix}?$$

   b Make up a matrix multiplication table for I, P, Q and H. What are the inverses of I, P, Q and H?

   c Repeat *a* and *b* for I, $R = \begin{pmatrix} 1 & 0 \\ 0 & -1 \end{pmatrix}$, $S = \begin{pmatrix} -1 & 0 \\ 0 & 1 \end{pmatrix}$ and H.

   d Do you notice any difference between the two tables?

Exercise 4

1 a In $\triangle ABC$ the equations of AB, BC and CA are $3x-2y+1=0$, $x+8y+9=0$ and $2x+3y-8=0$ respectively. Show by gradients that $\triangle ABC$ is right-angled at A.

b Calculate the coordinates of A, B and C, and show that AB:AC = 1:2.

2 a Find the equation of the circle which has centre $(2,3)$ and which passes through the point $(5,6)$.

b Prove that the line $x-y+7=0$ is a tangent to this circle.

c Show that where this line meets the circle $x^2+y^2=r^2$, x must satisfy the equation $2x^2+14x+(49-r^2)=0$.

d If the line $x-y+7=0$ is a tangent to the circle $x^2+y^2=r^2$, deduce the value of r.

3 a Write down an expression for the scalar product of $\boldsymbol{a}$ and $\boldsymbol{b}$: *(1)* in terms of the angle θ between them *(2)* in terms of their components. Deduce an expression for $\cos\theta$ in terms of the components.

b A is the point $(2,3-1)$, B is $(0,2,1)$ and C is $(-4,6,-3)$. Express the vectors represented by $\overrightarrow{AB}$ and $\overrightarrow{AC}$ in component form.

c Calculate the angle BAC, and the area of triangle ABC.

4 a Triangle PQR has vertices $P(-2,2)$, $Q(-1,4)$, $R(-2,4)$. Using matrices, or otherwise, find the coordinates of the vertices of the image triangle under the composite transformation of:

(1) reflection in the x-axis, followed by reflection in the line $y=x$ ($\triangle P_1Q_1R_1$)

(2) reflection in the line $y=x$, followed by reflection in the x-axis ($\triangle P_2Q_2R_2$)

b What single transformation would map $\triangle P_1Q_1R_1$ onto $\triangle P_2Q_2R_2$? If this single transformation is replaced by reflection in two axes, find the second axis, given that the first is:

(1) $y=0$ *(2)* $y=x$ *(3)* $y=2x$ *(4)* $y=mx\,(m\neq0)$

5 Find the equation of the image of the line $y=2x+3$ under a rotation about the origin of:

a π radians   *b* $\frac{1}{2}\pi$ radians   *c* $\frac{1}{4}\pi$ radians

Exercise 5

1   In $\triangle ABC$, A is the point $(4,4)$, B is $(1,-2)$ and C is $(6,3)$. P divides AB internally in the ratio $1:2$, Q divides AC internally in the ratio $3:1$ and R divides BC externally in the ratio $6:1$.

  a   Calculate the coordinates of P, Q and R and show that these points are collinear.

  b   In what ratio does Q divide PR?

2 *a*   Calculate the coordinates of the points in which the straight line $y = 2x - 3$ cuts the circle $x^2 + y^2 - 6x + 4y + 3 = 0$.

  b   If k is any real number verify that the circle
$$x^2 + y^2 - 6x + 4y + 3 + k(y - 2x + 3) = 0$$
must pass through the same two points.

  c   If this circle also passes through the point $(1,1)$, calculate the value of k. What are the centre and radius in this case?

3 *a*   If $\boldsymbol{u}$ has components $\frac{2}{3}, \frac{2}{3}$ and $-\frac{1}{3}$ show that $\boldsymbol{u}$ is a vector of unit length.

  b   If $\boldsymbol{v}$ has components $\frac{2}{3}, -\frac{1}{3}$ and $\frac{2}{3}$ show that $\boldsymbol{u}$ and $\boldsymbol{v}$ are perpendicular.

  c   Write down the components of a vector $\boldsymbol{w}$ of unit length which is perpendicular to both $\boldsymbol{u}$ and $\boldsymbol{v}$.

  d   If $p\boldsymbol{u} + q\boldsymbol{v} + r\boldsymbol{w} = \begin{pmatrix} 1 \\ 2 \\ 3 \end{pmatrix}$, find p, q and r.

4 *a*   The vertices of a square are $A(2,0)$, $B(4,0)$, $C(4,2)$ and $D(2,2)$. Find the image $A_1B_1C_1D_1$ of ABCD under the dilatation $[O, 5]$.

  b   Find also the image $A_2B_2C_2D_2$ of $A_1B_1C_1D_1$ under the mapping $(a, b) \to (a_1, b_1)$ where $a_1 = \frac{4}{5}a - \frac{3}{5}b$ and $b_1 = \frac{3}{5}a + \frac{4}{5}b$.

  c   Write down the matrices associated with the two transformations. Why can you conclude that these two transformations commute?

  d   Describe the composition of the two transformations geometrically.

5   O is the centre of two concentric circles of radii 1 and 2 units. A diameter PQORS of the larger circle cuts the smaller at Q and R. A chord PTV of the larger touches the smaller at T.

  a   Prove that triangles PTO and PVS are similar.

  b   Calculate the length of SV.

  c   Show that $\angle OPT = 30°$ and calculate $\angle TOS$.

  d   Show that $\triangle OVS$ is equilateral.

Exercise 6

1   A triangle has vertices A(6, 3), B(−4, 1) and C(4, −7). Find the equations of the medians AL and BM. Find the coordinates of G, the point of intersection of AL and BM.

Find the equation of CG, and show that CG produced bisects AB.

2 *a*   A point P(x, y) moves so that its distance from A(2, 1) is twice its distance from B(5, 1). Show that the locus of P is a circle, and find its centre and radius.

b   Write down the equation of AB. Verify that the circle cuts AB internally and externally in the ratio 2:1.

3 *a*   Under a transformation T, $(a, b) \rightarrow (a_1, b_1)$ so that $a_1 = 3a + 2b$ and $b_1 = 4a + 3b$. Write down the matrix associated with T and find its inverse.

Find the equation of the image of the line $x + 2y - 3 = 0$ under T.

b   If S is the transformation associated with the matrix $\begin{pmatrix} -10 & 15 \\ 15 & -20 \end{pmatrix}$, show by matrix multiplication that $T \circ S$ is equivalent to a dilatation [O, 5] followed by reflection in the line $y = x$.

4 *a*   A, B, C, D, E and F are six points in space with position vectors *a*, *b*, *c*, *d*, *e* and *f* respectively. Express the position vectors of K, L, M and N, the centroids of triangles ABC, ABD, DEF and CEF respectively in terms of *a*, *b*, *c*, *d*, *e* and *f*.

b   Show that K, L, M and N are vertices of a parallelogram.

5 *a*   Find the equation of the circle which passes through the points (−2, −2), (−1, 5) and (6, 4).

b   State its centre and radius.

6   A is the point (4, 3, −6), B is (2, 1, −4) and C is (−1, 4, −1).

a   Calculate the size of angle ABC.

b   D divides AB externally in the ratio 3:1. Find the coordinates of D, and show that CD is perpendicular to AB.

c   Illustrate in a diagram based on rectangular axes OX, OY and OZ.

Trigonometry

1 Products and Sums of Cosines and Sines

1 *Product formulae*

In Book 8 Trigonometry we studied the following Addition Formulae:

$$\cos(\alpha+\beta) = \cos\alpha\cos\beta - \sin\alpha\sin\beta \qquad . \qquad . \qquad (1)$$
$$\cos(\alpha-\beta) = \cos\alpha\cos\beta + \sin\alpha\sin\beta \qquad . \qquad . \qquad (2)$$
$$\sin(\alpha+\beta) = \sin\alpha\cos\beta + \cos\alpha\sin\beta \qquad . \qquad . \qquad (3)$$
$$\sin(\alpha-\beta) = \sin\alpha\cos\beta - \cos\alpha\sin\beta \qquad . \qquad . \qquad (4)$$

As we proved formula (1) for all α and β, and then derived the other formulae from (1), all four formulae are true for all α and β.

What can you say about the form of the right-hand sides of:

a all four formulae *b* formulae (1) and (2) *c* formulae (3) and (4)?

(i) Products of cosines, and products of sines

From formulae (1) and (2) we have, by addition,

$$\cos(\alpha+\beta) + \cos(\alpha-\beta) = 2\cos\alpha\cos\beta$$

We use this first in the form

$$2\cos\alpha\cos\beta = \cos(\alpha+\beta) + \cos(\alpha-\beta) \qquad . \qquad . \qquad (A)$$

This expresses a *product* of cosines as a *sum* of cosines.

Example 1. $2\cos 43° \cos 35° = \cos(43+35)° + \cos(43-35)°$
$$= \cos 78° + \cos 8°$$

Example 2. $2\cos 65° \cos 25° = \cos 90° + \cos 40°$
$$= 0 + \cos 40°$$
$$= \cos 40°$$

Example 3. $\cos 2\theta \cos \theta = \frac{1}{2}(\cos 3\theta + \cos \theta)$

189

Again from formulae (1) and (2), by subtraction,

$$\cos(\alpha - \beta) - \cos(\alpha + \beta)$$
$$= (\cos\alpha\cos\beta + \sin\alpha\sin\beta) - (\cos\alpha\cos\beta - \sin\alpha\sin\beta)$$
$$= 2\sin\alpha\sin\beta$$

We use this first in the form

$$2\sin\alpha\sin\beta = \cos(\alpha - \beta) - \cos(\alpha + \beta) \qquad . \qquad . \qquad (B)$$

This expresses a *product* of sines as a *difference* of cosines.

Example 4. $2\sin 27° \sin 14° = \cos(27-14)° - \cos(27+14)°$
$$= \cos 13° - \cos 41°$$

Example 5. $2\sin\frac{1}{3}\pi\sin\frac{1}{6}\pi = \cos\frac{1}{6}\pi - \cos\frac{1}{2}\pi$
$$= \tfrac{1}{2}\sqrt{3} - 0$$
$$= \tfrac{1}{2}\sqrt{3}$$

Notice in (A) and (B) that products of like functions are associated with cos (sum) and cos (difference). We may remember them in the form:

$$2\cos\ldots\cos\ldots = \cos(\text{sum}) + \cos(\text{difference}).$$
$$2\sin\ldots\sin\ldots = \cos(\text{difference}) - \cos(\text{sum}).$$

Exercise 1

Express each of the following as a sum of cosines:

1 $2\cos A\cos B$ *2* $2\cos x\cos y$ *3* $2\cos p\cos q$

4 $2\cos 50°\cos 30°$ *5* $2\cos 35°\cos 15°$ *6* $2\cos 53°\cos 13°$

Express each of the following as a difference of cosines:

7 $2\sin A\sin B$ *8* $2\sin x\sin y$ *9* $2\sin p\sin q$

10 $2\sin 60°\sin 20°$ *11* $2\sin 25°\sin 10°$ *12* $2\sin 35°\sin 15°$

Express as sums or differences of cosines, and simplify where possible:

13 $2\cos(x+y)\cos(x-y)$ *14* $2\sin(2x+y)\sin(2x-y)$

15 $2\cos\frac{1}{2}(\alpha+\beta)\cos\frac{1}{2}(\alpha-\beta)$ *16* $2\sin(A+B-C)\sin(A-B+C)$

17 $2\cos(\theta+\pi)\cos(\theta-\pi)$ *18* $2\sin(\theta+\frac{1}{4}\pi)\sin(\theta-\frac{1}{4}\pi)$

19	$\cos 50° \cos 30°$	*20*	$\cos 22° \cos 66°$	*21*	$\sin 130° \sin 20°$
22	$2 \cos 200° \cos 20°$	*23*	$2 \sin 75° \sin 15°$	*24*	$2 \cos \frac{3}{4}\pi \cos \frac{1}{4}\pi$

25 Show that $2 \sin\left(\frac{1}{4}\pi + \alpha\right) \sin\left(\frac{3}{4}\pi + \alpha\right) = \cos 2\alpha$

(ii) Products of cosines and sines

From formulae (3) and (4) we have, by addition,

$$\sin(\alpha + \beta) + \sin(\alpha - \beta) = 2 \sin \alpha \cos \beta$$

i.e. $\qquad 2 \sin \alpha \cos \beta = \sin(\alpha + \beta) + \sin(\alpha - \beta)$ . . (C)

And by subtraction, we have

$$\sin(\alpha + \beta) - \sin(\alpha - \beta) = 2 \cos \alpha \sin \beta$$

i.e. $\qquad 2 \cos \alpha \sin \beta = \sin(\alpha + \beta) - \sin(\alpha - \beta)$ . . (D)

(C) and (D) express *products* of cosines and sines as *sums* or *differences* of sines.

Example. Express $2 \sin 41° \cos 47°$ as a sum or difference of sines.
Using formula (C), $2 \sin 41° \cos 47° = \sin 88° + \sin(-6)°$
$$= \sin 88° - \sin 6°$$
Using formula (D), $2 \cos 47° \sin 41° = \sin 88° - \sin 6°$

Since the same result is obtained using both formulae, formula (C) by itself is sufficient. But it may be easier to remember both formulae in the form:
$$2 \sin \ldots \cos \ldots = \sin(\text{sum}) + \sin(\text{difference})$$
$$2 \cos \ldots \sin \ldots = \sin(\text{sum}) - \sin(\text{difference})$$

Notice that products of unlike functions are associated with $\sin(\text{sum})$ and $\sin(\text{difference})$.

Exercise 2

Express each of the following as a sum of sines:

1	$2 \sin A \cos B$	*2*	$2 \sin x \cos y$	*3*	$2 \sin p \cos q$
4	$2 \sin 50° \cos 30°$	*5*	$2 \sin 35° \cos 15°$	*6*	$2 \sin 60° \cos 24°$

Express each of the following as a difference of sines:

7	$2 \cos A \sin B$	*8*	$2 \cos x \sin y$	*9*	$2 \cos p \sin q$

10 $2\cos 60^\circ \sin 20^\circ$ **11** $2\cos 75^\circ \sin 5^\circ$ **12** $2\cos 25^\circ \sin 75^\circ$

Express as sums or differences of sines, and simplify where possible:

13 $\sin 5\theta \cos \theta$ **14** $\cos 3\theta \sin 2\theta$ **15** $\sin \tfrac{3}{2}\theta \cos \tfrac{1}{2}\theta$

16 $2\sin (P+Q)\cos (P-Q)$ **17** $2\sin (\theta+\tfrac{1}{2}\pi)\cos (\theta-\tfrac{1}{2}\pi)$

18 $2\sin \tfrac{1}{2}(\alpha+\beta)\cos \tfrac{1}{2}(\alpha-\beta)$ **19** $2\cos (\tfrac{1}{4}\pi+\alpha)\sin (\tfrac{1}{4}\pi-\alpha)$

20   Show that $4\sin 18^\circ \cos 36^\circ \sin 54^\circ = 1 + 2\sin 18^\circ - \cos 36^\circ$

Here, for reference, are the four product formulae:
$$2\cos \alpha \cos \beta = \cos (\alpha+\beta)+\cos (\alpha-\beta)$$
$$2\sin \alpha \sin \beta = \cos (\alpha-\beta)-\cos (\alpha+\beta)$$
$$2\sin \alpha \cos \beta = \sin (\alpha+\beta)+\sin (\alpha-\beta)$$
$$2\cos \alpha \sin \beta = \sin (\alpha+\beta)-\sin (\alpha-\beta)$$

Exercise 3B (Miscellaneous)

1   Write down the maximum and minimum values of:
  a $\sin x^\circ$ *b* $\cos x^\circ$ *c* $2\sin x^\circ$ *d* $3\cos x^\circ$ *e* $\sin 2x^\circ$

2   Express $2\cos (x+45)^\circ \cos (x-45)^\circ$ as a sum or difference, and hence find the maximum and minimum values of the product.

3   Repeat question **2** for the expressions:

  a $2\cos (x+30)^\circ \cos (x-30)^\circ$ *b* $2\sin (\theta+\tfrac{3}{4}\pi)\sin (\theta-\tfrac{3}{4}\pi)$

4   Simplify: $2\cos 50^\circ \cos 40^\circ - 2\sin 95^\circ \sin 85^\circ$.

5   Show that $2\sin 3\alpha \sin 4\alpha + 2\cos 5\alpha \cos 2\alpha - \cos 3\alpha = \cos \alpha$.

6   Show that
$$2\sin \tfrac{1}{2}\theta \cos \tfrac{3}{2}\theta + 2\sin \tfrac{5}{2}\theta \cos \tfrac{3}{2}\theta + 2\sin \tfrac{3}{2}\theta \cos \tfrac{7}{2}\theta = \sin 4\theta + \sin 5\theta.$$

7   Show that $\sin 3\beta + (\cos \beta + \sin \beta)(1 - 2\sin 2\beta) = \cos 3\beta$.

8   Without using tables, show that:
$$\sin 52^\circ \sin 68^\circ - \sin 47^\circ \cos 77^\circ - \cos 65^\circ \cos 81^\circ = \tfrac{1}{2}.$$

9   Express $\sin \alpha \sin 3\alpha$ as a difference of cosines. Hence find the sum of the first 6 terms of the series:
$$\sin \alpha \sin 3\alpha + \sin 2\alpha \sin 6\alpha + \sin 4\alpha \sin 12\alpha + \ldots$$

10   Find the sum of the first 6 terms of the series:
$$\cos 96\alpha \sin 32\alpha + \cos 48\alpha \sin 16\alpha + \cos 24\alpha \sin 8\alpha + \ldots$$

2   Sums and differences

From Section 1, $\cos(\alpha+\beta)+\cos(\alpha-\beta) = 2\cos\alpha\cos\beta$

$$\cos(\alpha-\beta)-\cos(\alpha+\beta) = 2\sin\alpha\sin\beta$$
$$\sin(\alpha+\beta)+\sin(\alpha-\beta) = 2\sin\alpha\cos\beta$$
$$\sin(\alpha+\beta)-\sin(\alpha-\beta) = 2\cos\alpha\sin\beta$$

Making the substitutions

$$\left.\begin{matrix} \alpha+\beta = C \\ \alpha-\beta = D \end{matrix}\right\} \text{ from which } \begin{cases} \alpha = \tfrac{1}{2}(C+D) \\ \beta = \tfrac{1}{2}(C-D) \end{cases}$$

we have

$$\cos C + \cos D = \quad 2\cos\tfrac{1}{2}(C+D)\cos\tfrac{1}{2}(C-D)$$
$$\cos C - \cos D = -2\sin\tfrac{1}{2}(C+D)\sin\tfrac{1}{2}(C-D)$$
$$\sin C + \sin D = \quad 2\sin\tfrac{1}{2}(C+D)\cos\tfrac{1}{2}(C-D)$$
$$\sin C - \sin D = \quad 2\cos\tfrac{1}{2}(C+D)\sin\tfrac{1}{2}(C-D)$$

In practice, the second of these formulae may be used as

$$\cos D - \cos C = 2\sin\tfrac{1}{2}(C+D)\sin\tfrac{1}{2}(C-D),\ \ C > D$$

Each of the formulae expresses a sum or difference of two cosines or of two sines as a product; the pattern persists that a sum or difference of cosines is a product of values of like functions, while a sum or difference of sines is a product of values of unlike functions.

We again emphasise that C and D are angles of any size by repeating the formulae in words:

$$\cos\ldots+\cos\ldots = \quad 2\cos(\text{half sum})\cos(\text{half difference})$$
$$\cos\ldots-\cos\ldots = -2\sin(\text{half sum})\sin(\text{half difference})$$
$$\sin\ldots+\sin\ldots = \quad 2\sin(\text{half sum})\cos(\text{half difference})$$
$$\sin\ldots-\sin\ldots = \quad 2\cos(\text{half sum})\sin(\text{half difference})$$

Example 1. $\sin 32° + \sin 28°$ *Example 2.* $\cos 5\theta - \cos 3\theta$

$\qquad\qquad = 2\sin 30°\cos 2°$ $\qquad\qquad\qquad\qquad = -2\sin 4\theta\sin\theta$

$\qquad\qquad = \cos 2°$

Exercise 4

Express the following as products, and simplify where possible:

1	$\cos A + \cos B$	2	$\cos 3X + \cos X$	3	$\cos 40° + \cos 10°$
4	$\cos P + \cos Q$	5	$\cos Y + \cos 5Y$	6	$\cos 80° + \cos 40°$
7	$\cos A - \cos B$	8	$\cos 4X - \cos 2X$	9	$\cos 50° - \cos 20°$
10	$\cos P - \cos Q$	11	$\cos Y - \cos 3Y$	12	$\cos 40° - \cos 20°$
13	$\sin A + \sin B$	14	$\sin 5X + \sin X$	15	$\sin 25° + \sin 15°$
16	$\sin P + \sin Q$	17	$\sin Y + \sin 3Y$	18	$\sin 170° + \sin 10°$
19	$\sin A - \sin B$	20	$\sin 3X - \sin X$	21	$\sin 44° - \sin 22°$
22	$\sin P - \sin Q$	23	$\sin 2Y - \sin 4Y$	24	$\sin 100° - \sin 80°$
25	$\cos 125° + \cos 55°$	26	$\sin 42° + \sin 18°$	27	$\cos 200° - \cos 20°$

28 $\sin(2\alpha + \beta) + \sin(2\alpha - \beta)$ 29 $\cos(\alpha - \beta) + \cos(\alpha + \beta)$

30 $\sin(x + h) - \sin x$ 31 $\cos(x + h) - \cos x$

32 $\sin(\tfrac{1}{2}\pi - \theta) + \sin(\tfrac{1}{2}\pi + \theta)$ 33 $\sin(\tfrac{1}{3}\pi + 2\theta) - \sin(\tfrac{1}{3}\pi - 2\theta)$

34 Show that

a $\dfrac{\sin 4\theta + \sin 2\theta}{\cos 4\theta + \cos 2\theta} = \tan 3\theta$ b $\dfrac{\cos 3\theta - \cos 5\theta}{\sin 3\theta - \sin \theta} = 2 \sin 2\theta$

35 If $x = \sin 3\theta + \sin \theta$ and $y = \cos 3\theta + \cos \theta$, show that:

a $x + y = 2 \cos \theta(\sin 2\theta + \cos 2\theta)$ b $x/y = \tan 2\theta$

c $x^2 + y^2 = 2 + 2 \cos 2\theta$

36a Prove that $\dfrac{\cos 2x° - \cos 4x°}{\sin 2x° \sin 3x°} = \dfrac{1}{\cos x°}$, $x \neq 0, 60, 90, 120, 180, \ldots$

b Deduce the solution set of the inequation $\dfrac{\cos 2x° - \cos 4x°}{\sin 2x° \sin 3x°} < 1$, for $0 < x < 60$.

37 If $\sin \alpha + \sin \beta = k$ and $\cos \alpha + \cos \beta = m$, prove that:

a $k + m = 2 \cos \tfrac{1}{2}(\alpha - \beta)[\sin \tfrac{1}{2}(\alpha + \beta) + \cos \tfrac{1}{2}(\alpha + \beta)]$
b $k = m \tan \tfrac{1}{2}(\alpha + \beta)$
c $k^2 + m^2 = 2[1 + \cos(\alpha - \beta)] = 4 \cos^2 \tfrac{1}{2}(\alpha - \beta)$.

3   Equations

(i) Multiple angles and general solution sets

Example 1. Solve the equation $\sin x° = 0·5$ in the cases where:

$$a \quad 0 \leqslant x < 360 \qquad b \quad x \in R.$$

a From tables, $\sin x° = 0·5 \Rightarrow x = 30$.

From Figure 1(i) or (ii), the other replacement for x in the domain $0 \leqslant x < 360$ is 150.

So for $0 \leqslant x < 360$, $x = 30$ or 150, i.e. the solution set is $\{30, 150\}$.

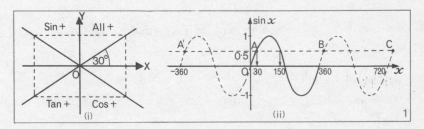

b The sine function has a period of 360, i.e. $\sin a° = \sin(360 + a)°$. So corresponding to the replacement at A there are others at B, C, ..., every 360, i.e. at $30 + 360$, $30 + 720$, ..., also at A', ..., i.e. at $30 - 360, ...$

All of these are given by $x = 30 + n.360$, $n \in Z$. So for $x \in R$, $x = 30 + n.360$, or $x = 150 + n.360$, i.e. the general solution set is $\{30 + n.360\} \cup \{150 + n.360\}$, $n \in Z$.

Example 2. Solve the equation $\sin 2x° = 0·5$ in the cases where:

$$a \quad 0 \leqslant x < 360 \qquad b \quad x \in R.$$

a From Example 1, $2x = 30, 30 + 360, ...$; or $150, 150 + 360, ...$
$$\qquad\qquad = 30, 390, ...; \text{ or } 150, 510, ...$$

So for $0 \leqslant x < 360$, $x = 15, 195; 75, 255$, i.e. the solution set is $\{15, 75, 195, 255\}$.

b For $x \in R$, $2x = 30 + n.360$, or $2x = 150 + n.360$, i.e. the general solution set is $\{15 + n.180\} \cup \{75 + n.180\}$, $n \in Z$.

Example 3. Solve $\tan \theta = -1$, where θ is in radian measure, for $0 \leqslant \theta < 2\pi$.

Now, $\tan \theta = 1 \quad \Rightarrow \quad \theta = \frac{1}{4}\pi$.

From Figure 1(i), $\tan \theta = -1 \quad \Rightarrow \quad \theta = \pi - \frac{1}{4}\pi$ or $2\pi - \frac{1}{4}\pi$.

So for $0 \leqslant \theta < 2\pi$, the solution set is $\{\frac{3}{4}\pi, \frac{7}{4}\pi\}$.

Exercise 5

Solve the following equations for $0 \leqslant x < 360$:

1	$\cos x° = 0·5$	2	$\sin x° = 1$	3	$\tan x° = 1$
4	$\cos x° = -0·707$	5	$\tan x° = 0·577$	6	$\sin x° = -0·5$
7	$\sin 2x° = 0·866$	8	$\cos 2x° = 0·5$	9	$\tan 2x° = 1·732$
10	$\sin 3x° = 1$	11	$\cos 3x° = -1$	12	$\sin 3x° = 0$

Solve the following equations for $0 \leqslant \theta < 2\pi$:

13	$\sin \theta = \frac{1}{\sqrt{2}}$	14	$\tan \theta = 0$	15	$\cos \theta = -\frac{1}{2}$.

Solve the following inequations for $0 \leqslant x < 360$:

16	$\sin x° > 0$	17	$\cos x° < 0$	18	$\sin x° \geqslant 0·5$

Exercise 5B

Give the general solutions for the equations in questions *1–6* of Exercise 5, i.e. for $x \in R$.

(ii) Equations involving sums of sines and cosines

Example 4. Solve $\cos x° - \cos 3x° + \sin x° = 0$, for $0 \leqslant x < 360$.

$$\cos x° - \cos 3x° + \sin x° = 0$$
$$\Leftrightarrow \quad 2 \sin 2x° \sin x° + \sin x° = 0$$
$$\Leftrightarrow \quad \sin x°(2 \sin 2x° + 1) = 0$$
$$\Leftrightarrow \quad \sin x° = 0 \text{ or } \sin 2x° = -0·5$$

$x = 0, 180$; or $2x = 210, 210 + 360$; $330, 330 + 360$.

i.e. $x = 105, 285$; $165, 345$.

For $0 \leqslant x < 360$, the solution set is $\{0, 105, 165, 180, 285, 345\}$.

Exercise 6

Solve the following equations for $0 \leqslant x < 360$:

1 $\sin 3x° + \sin x° = 0$ *2* $\cos 3x° + \cos x° = 0$

3 $\sin 3x° - \sin x° = 0$ *4* $\cos 3x° - \cos x° = 0$

5 $\cos 3x° + \cos x° + \cos 2x° = 0$ *6* $\sin 3x° + \sin x° = \sin 2x°$

7 $\cos x° - \cos 3x° + \sin 2x° = 0$ *8* $\sin 3x° - \sin x° = \cos 2x°$

9 $\sin 4x° + \sin 2x° = 0$ *10* $\cos 4x° + \cos 2x° = 0$

Solve the following equations for $0 \leqslant \theta < 2\pi$:

11 $\sin 4\theta - \sin 2\theta = 0$ *12* $\cos 4\theta - \cos 2\theta = 0$

13 $\cos 2\theta + \cos \theta = 0$ *14* $\sin 2\theta - \sin \theta = 0$

(iii) Quadratic equations in sin x and cos x (revision)

We saw in Book 8 that where an equation contains terms in $\sin 2x$ or $\cos 2x$, along with $\sin x$ or $\cos x$ or constants, we can form a quadratic equation in $\sin x$ or $\cos x$, using the formulae:

$$\sin 2x = 2\sin x \cos x; \quad \cos 2x = 2\cos^2 x - 1 = 1 - 2\sin^2 x.$$

Example 5. Solve $\cos 2x° + \cos x° - 2 = 0$ for $0 \leqslant x < 360$.

$$\cos 2x° + \cos x° - 2 = 0$$
$$\Leftrightarrow \quad 2\cos^2 x° - 1 + \cos x° - 2 = 0$$
$$\Leftrightarrow \quad 2\cos^2 x° + \cos x° - 3 = 0$$
$$\Leftrightarrow \quad (2\cos x° + 3)(\cos x° - 1) = 0$$
$$\Leftrightarrow \quad \cos x° = -1.5 \text{ or } \cos x° = 1$$

The solution set of $\cos x° = -1.5$ is ϕ; the solution set of $\cos x° = 1$ is $\{0\}$.

For $0 \leqslant x < 360$, the solution set is $\{0\}$.

Exercise 7

Solve the following equations for $0 \leqslant x < 360$:

1 $2\sin^2 x° - 7\sin x° + 3 = 0$ *2* $3\cos^2 x° + 2\cos x° - 1 = 0$

3 $\cos 2x° - \cos x° = 0$ *4* $\sin 2x° - \cos x° = 0$

5 $\cos 2x° - 3\cos x° + 2 = 0$ *6* $\cos 2x° - 3\sin x° - 1 = 0$

7    $\cos 2x° - 4\sin x° + 5 = 0$ 8 $5\cos 2x° + \cos x° + 2 = 0$

9    $3\sin 2x° + 5\cos x° = 0$ 10 $\cos 2x° + 2\sin x° = 1$

11    $3\cos 2x° - 10\cos x° + 7 = 0$ 12 $2\cos 2x° + 12\sin x° - 11 = 0$

Solve the following equations for $0 \leqslant \theta < 2\pi$:

13    $6\cos\theta + \cos 2\theta = 2\sin^2\theta - 5$ 14 $\cos 2\theta + 2\cos^2\theta = 2\sin\theta + 1$

Summary

1 *Product formulae*

$$2\cos\alpha\cos\beta = \cos(\alpha+\beta)+\cos(\alpha-\beta)$$
$$2\sin\alpha\sin\beta = \cos(\alpha-\beta)-\cos(\alpha+\beta)$$
$$2\sin\alpha\cos\beta = \sin(\alpha+\beta)+\sin(\alpha-\beta)$$
$$2\cos\alpha\sin\beta = \sin(\alpha+\beta)-\sin(\alpha-\beta)$$

2 *Sums and differences*

$$\cos C + \cos D = 2\cos\tfrac{1}{2}(C+D)\cos\tfrac{1}{2}(C-D)$$
$$\cos C - \cos D = -2\sin\tfrac{1}{2}(C+D)\sin\tfrac{1}{2}(C-D)$$
$$\sin C + \sin D = 2\sin\tfrac{1}{2}(C+D)\cos\tfrac{1}{2}(C-D)$$
$$\sin C - \sin D = 2\cos\tfrac{1}{2}(C+D)\sin\tfrac{1}{2}(C-D)$$

3 *Solution of equations*

Solutions for $0 \leqslant x < 360$, and for $x \in R$ (i.e. general solutions) can be obtained with the assistance of the following diagrams.

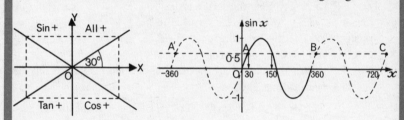

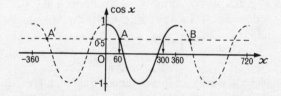

The Function f : x → a cos x + b sin x, and Applications

1 Introduction (for discussion)

Functions of the form
$f(x) = a\cos x + b\sin x$ occur frequently in mathematics and applied mathematics.

Suppose, for example, that in Figure 1 OP represents a crankshaft rotating about O at a constant 'angular velocity' of ω radians per second.

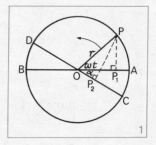

If P moves from an initial position A to P in t seconds, then the radian measure of $\angle AOP$ is ωt, and the position of P_1, the projection of P on OA is given by $OP_1 = r\cos\omega t$.

If P_2 is the projection of P on the diameter DOC such that $\angle COA = \alpha$ radians, then

$$OP_2 = r\cos(\omega t + \alpha) = r\cos\omega t\cos\alpha - r\sin\omega t\sin\alpha.$$

Since r and α are constants, this last expression is of the form $a\cos x + b\sin x$, where a and b are constants and $x \in R$.

The motions of P on the circle, of P_1 on AB and P_2 on CD are all periodic.

Practical illustrations

(i) When a coil is rotated in a magnetic field an alternating current is generated in the coil. The voltage of the current at time t is given by the formula $V = V_{max}\sin(\omega t + \alpha)$, where ω and α are constants similar to the ω, α referred to in connection with Figure 1, and V_{max} is the voltage induced at the instant when the rate of change of the number of 'lines of magnetic force' passing through the coil is greatest. The associated graph is shown in Figure 2.

200

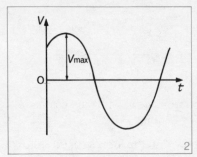
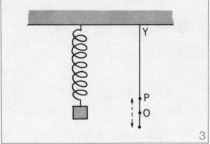

(ii) Another example is that of an object oscillating at the end of a spring, as shown in Figure 3. In applied mathematics it is proved that the acceleration of the object towards the midpoint O of the motion is given by an expression of the form $k \cos(\omega t + \alpha)$ or '$a \cos x + b \sin x$'.

In this chapter we study the function $a \cos x° + b \sin x°$, and we start by expressing it in the form $k \cos(x-\alpha)°$, i.e. in terms of one trigonometrical function. Similarly we express $a \cos x + b \sin x$ in the form $k \cos(x-\alpha)$.

2   *To express $a \cos x° + b \sin x°$ in the form $k\cos(x-\alpha)°$, where k is a positive constant and $0 < \alpha < 360$.*

Let $a \cos x° + b \sin x° = k \cos(x-\alpha)°$
$$= k \cos x° \cos \alpha° + k \sin x° \sin \alpha°.$$

Equating coefficients of $\cos x°$ and $\sin x°$, we have

$$\left. \begin{array}{l} k \cos \alpha° = a \\ k \sin \alpha° = b \end{array} \right\} \quad . \quad . \quad . \quad . \quad (1)$$

Squaring and adding, $k^2(\cos^2 \alpha° + \sin^2 \alpha°) = a^2 + b^2$, from which

$$k = \sqrt{(a^2 + b^2)}$$

From (1), $\tan \alpha° = \dfrac{b}{a}$. The quadrant of α is determined from the diagram by the signs of $\cos \alpha°$ and $\sin \alpha°$, which are those of a and b, since $k > 0$.

sine positive	all positive
tangent positive	cosine positive

Hence

$$a \cos x° + b \sin x° = k \cos (x - \alpha)°,$$

where $k = \sqrt{(a^2 + b^2)}$ and $\tan \alpha° = \dfrac{b}{a}$.

We base our work in this chapter on the formula for $\cos (x - \alpha)$, but any of $\cos (x \pm \alpha)$, $\sin (x \pm \alpha)$ may be used.

Example 1. Express $3 \cos x° + 4 \sin x°$ in the form $k \cos (x - \alpha)°$.

Let $3 \cos x° + 4 \sin x° = k \cos (x - \alpha)°$
$$= k \cos x° \cos \alpha° + k \sin x° \sin \alpha°$$

$\left. \begin{array}{l} k \cos \alpha° = 3 \\ k \sin \alpha° = 4 \end{array} \right\} \Rightarrow k = \sqrt{(3^2 + 4^2)} = 5$, and α is in the first quadrant.

Hence $\tan \alpha° = \frac{4}{3}$ and so $\alpha = 53 \cdot 1$.
It follows that $3 \cos x° + 4 \sin x° = 5 \cos (x - 53 \cdot 1)°$.

Example 2. Express $\cos \theta - \sin \theta$ in the form $k \cos (\theta - \alpha)$, where θ is in radian measure, and k and α are constants.

Let $\cos \theta - \sin \theta = k \cos (\theta - \alpha)$
$$= k \cos \theta \cos \alpha + k \sin \theta \sin \alpha$$

$\left. \begin{array}{l} k \cos \alpha = 1 \\ k \sin \alpha = -1 \end{array} \right\} \Rightarrow k = \sqrt{(1^2 + 1^2)} = \sqrt{2}$, and α is in the fourth quadrant.

Hence $\tan \alpha = -1$ and so $\alpha = \frac{7}{4}\pi$.
Hence $\cos \theta - \sin \theta = \sqrt{2} \cos (\theta - \frac{7}{4}\pi)$.

Alternatively, $\cos \theta - \sin \theta$ can be expressed in the form $k \cos (\theta + \alpha)$.
Let $\cos \theta - \sin \theta = k \cos (\theta + \alpha)$
$$= k \cos \theta \cos \alpha - k \sin \theta \sin \alpha$$

$\left. \begin{array}{l} k \cos \alpha = 1 \\ k \sin \alpha = 1 \end{array} \right\} \Rightarrow k = \sqrt{2}$, and α is in the first quadrant.

Hence $\tan \alpha = 1$ and so $\alpha = \frac{1}{4}\pi$.
Hence $\cos \theta - \sin \theta = \sqrt{2} \cos (\theta + \frac{1}{4}\pi)$.

Reminders

$\cos(x-\alpha)° = \cos x° \cos \alpha° + \sin x° \sin \alpha°$

π radians $= 180°$

Sine positive	All positive
Tangent positive	Cosine positive

Exercise 1

1 Assuming that $k > 0$, and $0 \leqslant \alpha < 360$, find k and α for each of the following pairs of equations:

a $k \cos \alpha° = 1, k \sin \alpha° = 1$ *b* $k \cos \alpha° = -1, k \sin \alpha° = 1$

c $k \cos \alpha° = 1, k \sin \alpha° = \sqrt{3}$ *d* $k \cos \alpha° = \sqrt{3}, k \sin \alpha° = -1$

e $k \cos \alpha° = 3, k \sin \alpha° = 4$ *f* $k \cos \alpha° = -5, k \sin \alpha° = 12$

2 Prove that: *a* $\cos x° - \sin x° = \sqrt{2} \cos(x-315)°$

$\qquad\qquad\quad$ *b* $\sqrt{3} \cos x° + \sin x° = 2 \cos(x-30)°$

$\qquad\qquad\quad$ *c* $\cos x° + 2 \sin x° = \sqrt{5} \cos(x-63\cdot4)°$

3 Express each of the following in the form $k \cos(x-\alpha)°$:

a $4 \cos x° + 3 \sin x°$ *b* $-\cos x° - \sin x°$ *c* $\cos x° - 3 \sin x°$

4 With θ in radian measure, express as $k \cos(\theta-\alpha)$:

a $\cos \theta + \sin \theta$ *b* $-\cos \theta + \sqrt{3} \sin \theta$

5 Express each of these in terms of a single trigonometrical function:

a $8 \cos x° + 6 \sin x°$ *b* $3 \cos x° - \sin x°$ *c* $\cos x° + \sqrt{3} \sin x°$

d $\sin x° - \cos x°$ *e* $8 \cos x° + 15 \sin x°$ *f* $-2 \cos x° - 2 \sin x°$

6 Using the expansion for $\cos(A-B)$, express $3 \cos \omega t + 4 \cos(\omega t - \frac{1}{3}\pi)$ in the form $a \cos \omega t + b \sin \omega t$.
Hence express $3 \cos \omega t + 4 \cos(\omega t - \frac{1}{3}\pi)$ in the form $R \cos(\omega t - \alpha)$, $R > 0$.

3   Solution of equations of the form
$a \cos x° + b \sin x° = c$

Example. Solve the equation $7 \cos x° + 4 \sin x° = 5$, for $0 \leqslant x < 360$.

Let $7 \cos x° + 4 \sin x° = k \cos (x - \alpha)°$
$$= k \cos x° \cos \alpha° + k \sin x° \sin \alpha°$$

$$\left.\begin{array}{l} k \cos \alpha° = 7 \\ k \sin \alpha° = 4 \end{array}\right\} \Rightarrow k = \sqrt{(4^2 + 7^2)} = \sqrt{65}, \text{ and } \alpha \text{ is in the first quadrant.}$$

Hence $\tan \alpha° = \frac{4}{7}$ and so $\alpha = 29 \cdot 7$.

The equation is $\sqrt{65} \cos (x - 29 \cdot 7)° = 5$

$$\Rightarrow \cos (x - 29 \cdot 7)° = \frac{5}{\sqrt{65}}$$

$$\Rightarrow x - 29 \cdot 7 = 51 \cdot 6 \text{ or } 308 \cdot 4$$

$$\Rightarrow x = 81 \cdot 3 \text{ or } 338 \cdot 1$$

nos	logs
5	0·699
$\sqrt{65}$	2 | 1·813
	0·906
$\cos 51 \cdot 6°$	$\bar{1} \cdot 793$

Exercise 2

Solve the equations in questions *1–8* for $0 \leqslant x < 360$:

1   $3 \cos x° + 4 \sin x° = 5$ 2 $5 \cos x° - 12 \sin x° = 13$

3   $2 \cos x° + \sqrt{5} \sin x° = 1 \cdot 5$ 4 $2 \cos x° - \sin x° = \sqrt{5}$

5   $\sin x° + \cos x° = -1$ 6 $2 \sin x° + 2 \cos x° = 1$

7   $\cos x° - \sqrt{3} \sin x° - 1 = 0$ 8 $9 \cos x° + 40 \sin x° = 0$

9   Express each of the following in the form $k \sin (x - a)°$. Then solve the equation $f(x) = 1$.

a $f(x) = \sin x° - \sqrt{3} \cos x°$ b $f(x) = 6 \sin x° - 8 \cos x°$

10 In each of the following equations, θ is in radian measure. Solve the equations for $0 \leqslant \theta < 2\pi$:

a $\sqrt{3} \cos \theta + \sin \theta = 1$ b $\cos \theta - \sin \theta = -1$

11   Express $3\cos 2x° + 4\sin 2x°$ in the form $k\cos(2x-\alpha)°$, and hence solve the equation $3\cos 2x° + 4\sin 2x° = 5, 0 \leqslant x < 180$.

12   Express $8\cos 2x° - 6\sin 2x°$ in the form $k\cos(2x-\alpha)°$, and hence solve the equation $8\cos 2x° - 6\sin 2x° = 5, 0 \leqslant x < 180$.

4 Maximum and minimum values; graph sketching

We can find the approximate maximum and minimum values of the function f defined by $f: x \rightarrow \cos x° + \sin x°$, $0 \leqslant x < 360$, from the graph of $y = f(x)$. The graph can be drawn by adding corresponding ordinates of the graphs of $y = \cos x°$ and $y = \sin x°$, as shown in Figure 5.

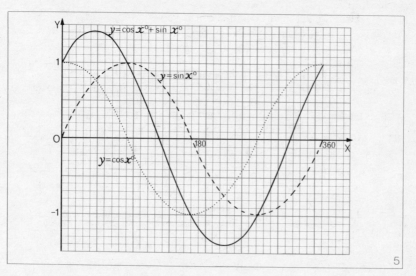

From the graph it appears that for $0 \leqslant x < 360$, the maximum and minimum values of f are approximately $1\cdot4$ and $-1\cdot4$, and the zeros of f are $x = 135$ and $x = 315$.

By expressing $\cos x° + \sin x°$ in the form $k\cos(x-\alpha)°$, we can find the maximum and minimum values of f more quickly and accurately as follows:

Let $\cos x° + \sin x° = k \cos(x-a)°$
$$= k \cos x° \cos \alpha° + k \sin x° \sin \alpha°$$

$\left.\begin{array}{l} k \cos \alpha° = 1 \\ k \sin \alpha° = 1 \end{array}\right\} \implies k = \sqrt{2}$, and α is in the first quadrant.

Hence $\tan \alpha° = 1$ and so $\alpha = 45$.
Hence $\cos x° + \sin x° = \sqrt{2} \cos(x-45)°$.

Referring to Figure 6(i), we recall that, for $0 \leqslant x < 360$,

(i) the maximum value of $\cos x°$ is 1, when $x = 0$
(ii) the minimum value of $\cos x°$ is -1, when $x = 180$.

So the maximum value of f is $\sqrt{2}$, and this occurs when $\cos(x-45)° = 1$, i.e. when $x = 45$ $(0 \leqslant x < 360.)$

The minimum value of f is $-\sqrt{2}$. This occurs when $\cos(x-45)° = -1$, i.e. when $x-45 = 180$, i.e. $x = 225$, in the interval $0 \leqslant x < 360$.

Also, the zeros of f occur when $\sqrt{2} \cos(x-45)° = 0$
i.e. when $x-45 = 90$ or 270
i.e. when $x = 135$ or 315 $(0 \leqslant x < 360)$.

For interest, Figure 6 shows the graphs of (i) $y = \cos x°$ (ii) $y = \cos(x-45)°$ (iii) $y = \sqrt{2} \cos(x-45)°$. (ii) is the image of (i) under the translation $\begin{pmatrix} 45 \\ 0 \end{pmatrix}$; (iii) is obtained from (ii) by multiplying the values of the function by $\sqrt{2}$.

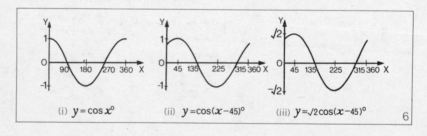

(i) $y = \cos x°$ (ii) $y = \cos(x-45)°$ (iii) $y = \sqrt{2}\cos(x-45)°$

6

Exercise 3

1   Write down the maximum and minimum values of the following functions f, and find the corresponding replacements for x $(0 \leqslant x < 360)$.

a $f(x) = \cos x°$ b $f(x) = 2 \cos (x - 30)°$ c $f(x) = 5 \cos (x - 200)°$

For each of the functions in questions 2–5:

a   express the function in the form $k \cos (x - \alpha)°$.
b   obtain the maximum and minimum values of the function, and the corresponding replacements for x $(0 \leqslant x < 360)$
c   find the zeros of the function
d   sketch the graph of the function.

2   $f(x) = 3 \cos x° + 4 \sin x°$ 3 $f(x) = \cos x° + \sqrt{3} \sin x°$

4   $f(x) = 3 \cos x° - 4 \sin x°$ 5 $f(x) = 7 \cos x° - 24 \sin x°$

6   Show that the function f given by $f(x) = 3 \cos x° + 4 \sin x° - 5$ has maximum value 0 and minimum value -10. Sketch its graph.

7   Show that the function $f : x \to \sqrt{3} \sin x° - \cos x° + 2$ has maximum value 4 and minimum value 0. Sketch the graph of the function.

8   Given that $0 < \theta < \frac{1}{2}\pi$, and $Q = \dfrac{2c}{\sqrt{3} \cos \theta + \sin \theta}$, c being a positive constant, find the minimum value of Q.

9   A right-angled triangle has an angle of $x°$ and a hypotenuse h cm long. If the perimeter of the triangle is 10 cm, show that

$$h = \frac{10}{1 + \cos x° + \sin x°}.$$

By first expressing $\cos x° + \sin x°$ in the form $k \cos (x - \alpha)°$, show that the minimum value of h is 4·1 cm, to two significant figures.

10   The formula $h = 38 \, (2 \sin 30t° + \cos 30t°)$ gives the height in centimetres of the sea above the mean level t hours after midnight, at a certain port. Express h in the form $k \cos (30t - \alpha)°$, and find the times of high and low water during the first twenty-four hours, and also the times at which the tide is 50 cm above the mean level.

11 In Figure 7, line OP of length 3 cm makes angle $a°$ with the x-axis. PQ, of length 2 cm, is perpendicular to OP. M and N are the projections of Q and P on the x-axis.

a Show that angle MQP = $a°$.

b Show that MQ = $2\cos a° + 3\sin a°$, and hence express MQ in the form $k\cos(x-\alpha)°$. Obtain a similar expression for OM.

c Deduce the maximum values of OM and MQ.

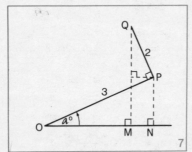

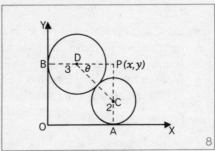

12 Figure 8 shows two circles, centres C and D, of radii 2 m and 3 m respectively. The circles touch each other externally, and touch OX and OY at A and B. AC and BD produced meet at P(x, y), and angle CDP = θ radians. Show that:

a $x = 3 + 5\cos\theta$ b $y = 2 + 5\sin\theta$

Hence express AB^2 in the form $p + q\cos\theta + r\sin\theta$, stating the values of p, q and r. Find the maximum length of AB.

Summary

1 Reminders

(i) $\cos(x-\alpha)° = \cos x° \cos \alpha° + \sin x° \sin \alpha°$

(ii)

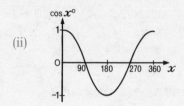

2 $a \cos x° + b \sin x° = k \cos(x-\alpha)°,$

where $k = \sqrt{(a^2+b^2)}$ and $\tan \alpha° = \dfrac{b}{a}$, and the quadrant for α is the quadrant of the point (a, b).

3   *Maximum and minimum values of $a \cos x° + b \sin x°$* are given by the maximum and minimum values of $k \cos(x-\alpha)°$. These are:

(i) k, when $\cos(x-\alpha)° = 1$, i.e. $x-\alpha = 0$, or $x = \alpha$

(ii) $-k$, when $\cos(x-\alpha)° = -1$, i.e. $x-\alpha = 180$, or $x = \alpha + 180$

4   *Solutions of the equation $a \cos x° + b \sin x° = c$* are obtained from the equation $k \cos(x-\alpha)° = c$, i.e. $\cos(x-\alpha)° = \dfrac{c}{k}$.

Revision Exercises

Revision Exercise on Chapter 1
Products and Sums of Cosines and Sines

Revision Exercise 1

Express each of the following as a sum or difference, and simplify where possible (questions *1–4*):

1 a $2\cos X \cos Y$ *b* $2\cos 50° \cos 30°$ *c* $2\sin X \sin Y$

   d $2\cos\frac{1}{2}(\alpha+\beta)\cos\frac{1}{2}(\alpha-\beta)$ *e* $2\sin\frac{1}{2}(p+q)\sin\frac{1}{2}(p-q)$

2 a $2\sin X \cos Y$ *b* $2\sin 40° \cos 30°$ *c* $2\cos R \sin S$

   d $2\cos\frac{1}{2}(\alpha+\beta)\sin\frac{1}{2}(\alpha-\beta)$ *e* $2\sin(x+\theta)\cos(x-\theta)$

3 a $2\cos 75° \cos 15°$ *b* $2\sin 40° \cos 20°$ *c* $2\cos 25° \sin 25°$

   d $2\sin\frac{3}{4}x° \sin\frac{1}{4}x°$ *e* $\cos 2a° \cos a°$ *f* $\sin 4b° \sin 2b°$

4 a $2\cos(\pi+3\theta)\cos(\pi-\theta)$ *b* $\sin(45+a)° \cos(45-a)°$

   c $2\sin(\theta-\frac{1}{6}\pi)\cos(\theta+\frac{1}{6}\pi)$ *d* $2\sin(x+105)° \cos(x+75)°$

5   Write down the maximum and minimum values of:

   a $\sin x°$ *b* $5\sin x°$ *c* $\sin 5x°$ *d* $\cos x°$ *e* $-\cos x°$

6   Express each of the following as a sum or difference, and hence find the maximum and minimum values of the product:

   a $2\cos(x+20)° \cos(x-10)°$ *b* $2\sin(\frac{1}{6}\pi+\theta)\sin(\frac{1}{6}\pi-\theta)$

   Express each of the following as a product, and simplify where possible (questions *7–8*):

7 a $\cos 60° + \cos 20°$ *b* $\sin 80° + \sin 60°$ *c* $\sin 10° - \sin 20°$

   d $\cos 50° - \cos 130°$ *e* $\sin\frac{5}{4}x° + \sin\frac{3}{4}x°$ *f* $\cos 7\theta - \cos 3\theta$

210

8 *a* $\sin(\theta+h)-\sin\theta$ *b* $\cos(\theta+h)-\cos\theta$

   c $\cos(\alpha-\tfrac{1}{6}\pi)+\cos(\alpha+\tfrac{1}{6}\pi)$ *d* $\sin(\alpha+\tfrac{1}{2}\pi)+\sin(\alpha-\tfrac{1}{2}\pi)$

9   Simplify: *a* $\dfrac{\cos 75°+\cos 15°}{\sin 75°-\sin 15°}$ *b* $\dfrac{\cos 5\theta-\cos 3\theta}{\sin 5\theta+\sin 3\theta}$

10   Express $x=\sin 4\theta+\sin 2\theta$ and $y=\cos 4\theta+\cos 2\theta$ as products.

   Hence or otherwise show that: *a* $x/y=\tan 3\theta$

                                    b $x^2+y^2=2+2\cos 2\theta$

11*a* Prove that $\cos 6x+\cos 2x-\cos 4x=\cos 4x(2\cos 2x-1)$.

   b Hence prove that $\dfrac{\sin 6x+\sin 2x-\sin 4x}{\cos 6x+\cos 2x-\cos 4x}=\tan 4x$

   Solve the following equations for $0\leqslant x<360$ (questions *12–14*):

12*a* $\tan x°=\surd 3$ *b* $\cos 2x°=-1$ *c* $\sin 3x°=-0{\cdot}707$

13*a* $\cos 4x°+\cos 2x°=0$ *b* $\cos 3x°+\cos x°-2\cos 2x°=0$

   c $\sin x°+\sin 5x°=\sin 3x°$ *d* $\sin 4x°-\sin 2x°=\surd 2\sin x°$

14*a* $\sin 2x°+\cos x°=0$ *b* $\cos 2x°-\sin x°=0$

   c $2\sin^2 x°-5\sin x°+3=0$ *d* $\cos 2x°+\cos x°=0$

15   Prove that $\sin a°+\sin(a+120)°+\sin(a+240)°=0$

16   Show that $\cos(\tfrac{1}{4}\pi+\theta)\cos(\tfrac{1}{4}\pi-\theta)=\tfrac{1}{2}\cos 2\theta$

17   Find the maximum and minimum values of the following, for $0\leqslant x\leqslant 180$, and give the corresponding replacements for x:

   a $\sin(x+20)°\cos(x-10)°$ *b* $\cos(x-30)°\cos(x+60)°$

18   Prove that $\dfrac{\cos 3x-\sin 6x-\cos 9x}{\sin 9x-\cos 6x-\sin 3x}=\tan 6x$

19   Solve the following equations for $0\leqslant x<360$:

   a $\cos x°+\cos 2x°+\cos 3x°=0$

   b $\cos x°+\cos 2x°=\cos 4x°+\cos 5x°$

20 If $V=200\sin 360t$, $I=3{\cdot}5\sin(360t+c)$ and $W=IV$, show that
    $W=350[\cos c-\cos(720t+c)]$.

21a Express $2\sin(\alpha+\beta)\sin(\alpha-\beta)$ as the difference of two cosines. Hence show that $\sin(\alpha+\beta)\sin(\alpha-\beta) = \sin^2\alpha - \sin^2\beta$.

b Deduce that $\sin 8\theta \sin 6\theta = \sin^2 7\theta - \sin^2 \theta$. Prove this result also by factorising the right side.

Revision Exercise on Chapter 2
The function f: x → a cos x + b sin x

Revision Exercise 2

1 Assuming that $k > 0$ and $0 \leqslant \alpha < 360$, find k and α for each of the following:

a $k\cos\alpha° = 1, k\sin\alpha° = 1$ *b* $k\cos\alpha° = 1, k\sin\alpha° = -1$

c $k\cos\alpha° = \sqrt{3}, k\sin\alpha° = 1$ *d* $k\cos\alpha° = -\sqrt{3}, k\sin\alpha° = 1$

e $k\cos\alpha° = 8, k\sin\alpha° = 6$ *f* $k\cos\alpha° = -4, k\sin\alpha° = -3$

2 Prove that: *a* $2\cos x° + 2\sin x° = 2\sqrt{2}\cos(x-45)°$

               b $\sin\theta - \sqrt{3}\cos\theta = 2\cos(\theta - \tfrac{5}{6}\pi)$, θ in radians.

3 Express each of the following in the form $k\cos(x-\alpha)°$:

a $7\cos x° - 5\sin x°$ *b* $-3\cos x° - 2\sin x°$

4 Express the following in the form $k\cos(\theta-\alpha)$, θ and α in radians:

a $4\cos\theta - 4\sin\theta$ *b* $\sqrt{3}\cos\theta - \sin\theta$

5 Express the following in terms of single trigonometrical functions:

a $6\cos x° - 7\sin x°$ *b* $4\cos x° + 5\sin x°$

c $3\sin x° - 5\cos x°$ *d* $9\cos x° + 2\sin x°$

6 Write down the maximum and minimum values of the expressions in question *5*.

     In questions *7* and *8* calculate the maximum and minimum values of the functions *f*, and the corresponding replacements for x in the domain $0 \leqslant x < 360$. Find the zeros of the functions, and sketch the graphs of the functions.

7 $f(x) = \sin x° - \sqrt{3}\cos x°$ *8* $f(x) = 6\cos x° + 8\sin x°$

Find the maximum and minimum values of the functions in questions *9* and *10*.

9 $f: x \to 12 \cos x° + 5 \sin x° - 12$ *10* $f: x \to \dfrac{1}{4 + \sqrt{3} \sin x° + \cos x°}$

Solve the following equations for $0 \leqslant x < 360$:

11 $\sin x° + \sqrt{3} \cos x° = 1$ *12* $3 \cos 2x° + 4 \sin 2x° = 5$

13 Express $f(x) = a \cos x + b \sin x$ in the form $k \cos(x - \alpha)$.
Explain why the equation $f(x) = c$ has real roots if and only if $c^2 \leqslant a^2 + b^2$, i.e. $|c| \leqslant \sqrt{(a^2 + b^2)}$.
   Find the solution sets of the equations of the form $a \cos x + b \sin x = c$, given by:

a $a = 1, b = 1, c = \sqrt{3}, x \in R$

b $a = 1, b = -1, c = \sqrt{2}, -\tfrac{1}{2}\pi < x < \tfrac{1}{2}\pi$

c $a = \sqrt{3}, b = -1, c = 2, 0 \leqslant x < 2\pi$

d $a = 2, b = -3, c = 4, 0 \leqslant x < 4\pi$

14 The displacement, x centimetres, of a point on a sliding mechanism from a fixed point O after t seconds is given by $x = 8 \sin 2t - 15 \cos 2t$. By expressing x in the form $A \sin(2t - \alpha)$, find:

a the maximum displacement

b the least value of t for which the point is at maximum positive displacement. (Note that the angles $2t$ and α are in radians.)

Cumulative
Revision
Section

Cumulative Revision Section (Books 5-9)

Revision Topic 1
Angles, Triangles and Three Dimensions

Reminders

1   *Definitions.* P has Cartesian coordinates (x, y) and polar coordinates $(r, a°)$.

$\cos a° = \dfrac{x}{r}$, $\sin a° = \dfrac{y}{r}$, $\tan a° = \dfrac{y}{x}$;   P is the point $(r\cos a°, r\sin a°)$

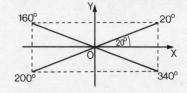

2   *Special angles.*

$\cos 45° = \frac{1}{\sqrt{2}}$, $\tan 45° = 1$, $\sin 30° = \frac{1}{2}$, $\cos 60° = \frac{1}{2}$, etc.

3   *Extending the use of the trigonometrical tables: $a > 90$.*

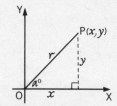

sin +	all +
tan +	cos +

e.g. $\sin 160° = \sin 20° = 0.342$
$\cos 200° = -\cos 20° = -0.940$
$\tan 340° = -\tan 20° = -0.364$

4   *The graphs of the sine and cosine functions*

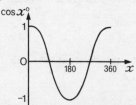

Period of graphs $= 360°$; $\sin 90° = 1$, $\sin 180° = 0$, $\cos 180° = -1$, etc.

5   *Formulae.*   For all A, $\cos^2 A + \sin^2 A = 1$ and $\tan A = \dfrac{\sin A}{\cos A}$

6   For every triangle ABC we have:

a   *The Sine Rule:* $\dfrac{a}{\sin A} = \dfrac{b}{\sin B} = \dfrac{c}{\sin C}$

b   *The Cosine Rule:* $a^2 = b^2 + c^2 - 2bc \cos A$, or $\cos A = \dfrac{b^2 + c^2 - a^2}{2bc}$

c   *The Area of a Triangle:* $\triangle = \frac{1}{2}bc \sin A$

Exercise 1

1   Which of the following are true and which are false, for $0 < x < 360$, within the accuracy of three-figure tables?

a $\sin x° = 0{\cdot}884 \ \Rightarrow \ x = 62{\cdot}1$ or $242{\cdot}1$
b $\tan y° = 0{\cdot}596 \ \Rightarrow \ y = 30{\cdot}8$ or $210{\cdot}8$
c $\cos z° = -0{\cdot}772 \ \Rightarrow \ z = 140{\cdot}5$ or $219{\cdot}5$

2   Find the Cartesian coordinates of the point P, given that:

a $OP = 3$, and $\angle XOP = 53{\cdot}1°$
b $OP = 5$, and $\angle XOP = 341{\cdot}8°$

3   Without using tables find the exact values of:

a $\sin x°$ and $\tan x°$, if $\cos x° = \frac{5}{13}$ and $0 < x < 90$
b $\sin a°$ and $\cos a°$, if $\tan a° = -\frac{9}{40}$ and $0 < a < 180$

4   Without using tables give the exact values of:

a $\sin 30°$ b $\cos 45°$ c $\cos 135°$ d $\tan 60°$ e $\tan 300°$

5   Solve for $0 \leqslant x \leqslant 360$:   a $2 \sin x + 1 = 0$ b $2 \cos x - \sqrt{3} = 0$

6   Find a replacement for $x, 0 \leqslant x \leqslant 180$, for which the sine and cosine functions have equal values. Illustrate by sketching their graphs.

7   A triangle has sides of lengths 3 cm, 5 cm and 7 cm. Show that its greatest angle is $120°$, and calculate its area.

8   Calculate the smallest angle of a triangle with sides of lengths 4 km, 6 km and 8 km.

9   The greatest side of a triangle is 14·5 cm long. Two of the angles are $45°$ and $60°$. Find the other sides and the area of the triangle.

10   Given $a = 4$ cm and $b = 5$ cm, and the area of triangle ABC = 7·85 cm^2, calculate the possible sizes of angle C.

11   In triangle ABC, A = 51°, B = 68°, $c = 4·5$ m. Calculate:

  a b b the area of triangle ABC.

12   In triangle ABC, $a = 9·5$ cm, $b = 8·4$ cm, C = 73°. Calculate:

  a the area of the triangle   b c c A and B.

13   An aircraft leaves airfield F and flies 120 kilometres on course 032° to a point G. It then changes course to 143° and flies 90 kilometres to a point H. Find the distance and bearing of H from F.

14   A cuboid has base ABCD and top EFGH; AE, BF, CG and DH are edges. AB = 5 cm, BC = 12 cm, CG = 4 cm. Find:

  a the lengths of ED and AG
  b the angles which HC and EC make with plane ABCD
  c the angle between the planes EBCH and ABCD.

15   Figure 1 shows a cuboid. Calculate:

  a the coordinates of T, S and V
  b the angle between OV and plane OPQR
  c the angle between OU and plane OPQR
  d the coordinates of the point of intersection of the space diagonals
  e the angle between VP and OU
  f the angle between planes QRST and OPTS.

16   V,ABCD is a pyramid on a square base of side 1·6 m. Each slant edge is 2·2 m long. Calculate:

  a the height of the pyramid
  b the angle each sloping face makes with the base.

17   Triangle ABC, right-angled at A, has AB and AC each of length 8 cm. AP is perpendicular to the plane ABC and is of length 15 cm. PB and PC are drawn. Calculate:

  a ∠BPC   b the angle between planes ABC and PBC.

Revision Topic 2   The Addition Formulae

Reminders

1   *Related angles*

$$\left.\begin{array}{l}\cos(180-a)^\circ = -\cos a^\circ \\ \sin(180-a)^\circ = \sin a^\circ\end{array}\right\} \qquad \left.\begin{array}{l}\cos(-a)^\circ = \cos a^\circ \\ \sin(-a)^\circ = -\sin a^\circ\end{array}\right\}$$

$$\left.\begin{array}{l}\cos(90-a)^\circ = \sin a^\circ \\ \sin(90-a)^\circ = \cos a^\circ\end{array}\right\}$$

2   *Radian measure*

Radian measure of $\angle$ AOB at centre of circle $= \dfrac{\text{arc AB}}{\text{radius OA}}$.

$180^\circ = \pi$ radians;   1 radian $\doteqdot 57^\circ$; $\sin \pi = \sin 180^\circ$

3   *Addition formulae*

$\cos(\alpha+\beta) = \cos\alpha\cos\beta - \sin\alpha\sin\beta$
$\cos(\alpha-\beta) = \cos\alpha\cos\beta + \sin\alpha\sin\beta$
$\sin(\alpha+\beta) = \sin\alpha\cos\beta + \cos\alpha\sin\beta$
$\sin(\alpha-\beta) = \sin\alpha\cos\beta - \cos\alpha\sin\beta$

$\tan(\alpha+\beta) = \dfrac{\tan\alpha + \tan\beta}{1 - \tan\alpha\tan\beta}$

$\tan(\alpha-\beta) = \dfrac{\tan\alpha - \tan\beta}{1 + \tan\alpha\tan\beta}$

4   *Double angle formulae*

$\sin 2\alpha = 2\sin\alpha\cos\alpha$
$\cos 2\alpha = \cos^2\alpha - \sin^2\alpha$
$\quad\quad = 2\cos^2\alpha - 1$
$\quad\quad = 1 - 2\sin^2\alpha$
$\cos^2\alpha = \frac{1}{2}(1 + \cos 2\alpha)$
$\sin^2\alpha = \frac{1}{2}(1 - \cos 2\alpha)$

$\tan 2\alpha = \dfrac{2\tan\alpha}{1 - \tan^2\alpha}$

Exercise 2

1   Find the values of:

a $\sin 200^\circ$ b $\cos 200^\circ$ c $\tan 200^\circ$ d $\tan 250{\cdot}5^\circ$

2   If $\sin a^\circ = k$, express the following in terms of k:

a $\sin(180-a)^\circ$ b $\sin(180+a)^\circ$ c $\sin(-a)^\circ$ d $\sin(360-a)^\circ$

3 a   Express in radian measure: $30^\circ, 45^\circ, 120^\circ, 240^\circ, 300^\circ, 360^\circ$.
  b   Express in degree measure: $\frac{1}{3}\pi, \frac{1}{2}\pi, \frac{3}{2}\pi, \frac{5}{6}\pi, \frac{7}{12}\pi$ radians.

4   A radar scanner rotates at a speed of 24 revolutions per minute. Calculate its angular velocity in:

a   degrees per second   b   radians per second.

5   Calculate the length of arc of a circle of radius 18 cm subtended by an angle of 80° at the centre of the circle. (Take 3·14 for π.)

6   Solve for $0 < x < 720$:

   a $2 \sin x° = 1$ *b* $\tan x° - 1 = 0$ *c* $\sqrt{2} \cos x° + 1 = 0$

7   Prove:

   a $\sin(A+45)° + \sin(A-45)° = \sqrt{2} \sin A°.$
   b $\sin(\theta + \frac{7}{4}\pi) = \frac{1}{\sqrt{2}}(\sin\theta - \cos\theta)$

8   If $\sin\alpha = \frac{3}{5}$ and $\cos\beta = \frac{12}{13}$, where α and β are acute angles, find, without using tables, the values of $\cos(\alpha+\beta)$ and $\cos 2\alpha$.

9 *a* If $\tan A = 2$ and $\tan B = 3$, find, without using tables, the value of $\tan(A+B)$.

   b If $\tan P = \dfrac{n}{n-1}$ and $\tan Q = \dfrac{1}{2n-1}$. $n \neq \frac{1}{2}$ or 1, find $\tan(P-Q)$ in in its simplest form.

10   Express:

   a $\cos^2\theta$ in terms of $\cos 2\theta$ *b* $\cos^2\frac{1}{2}x$ in terms of $\cos x$

11   Simplify:

   a $\sin 25° \cos 65° + \cos 25° \sin 65°$ *b* $6 \sin x \cos x$

   c $\cos 100° \cos 10° + \sin 100° \sin 10°$ *d* $2 \cos^2 4x - 1$

12   Solve for $0 \leqslant x \leqslant 360$:

   a $\cos 2x° + \cos x° + 1 = 0$ *b* $\cos 2x° + \sin x° - 1 = 0$

   c $\sin 2x° - \sin x° = 0$ *d* $\tan 2x° = 3 \tan x°$

13   P is (5, 12) and Q is (8, 6). Use the $\sin(\alpha-\beta)$ formula to calculate the value of $\sin \angle QOP$.

14   R is $(4, -3)$ and S is $(-5, 12)$. Calculate the value of $\cos \angle ROS$.

15   State whether each of the following is true or false.

   a $\cos\pi = -1$
   b The minimum value of $\sin^2 x°$ is 0.
   c $\cos(A+B) = \cos A + \cos B$
   d The period of the graph $y = \sin 4x°$ is 90.

16 If $p = \sin x - \sin y$ and $q = \cos x - \cos y$, show that
$$p^2 + q^2 = 2[1 - \cos(x - y)].$$

17 Prove the following identities:

   a $\cos^4 \alpha - \sin^4 \alpha = \cos 2\alpha$

   b $\tan(x + 45)° \tan(x - 45)° = -1$

   c $\cos A - \sin A \sin 2A = \cos A \cos 2A$

   d For $0 < \theta < \pi$,

$$\sqrt{\left(\frac{1 - \cos\theta}{1 + \cos\theta}\right)} = \tan \tfrac{1}{2}\theta. \quad (Hint: \cos 2x = 2\cos^2 x - 1 = 1 - 2\sin^2 x)$$

18a Expressing $\cos 3\theta$ as $\cos(2\theta + \theta)$, show that $\cos 3\theta = 4\cos^3 \theta - 3\cos\theta$.

   b Find a formula for $\sin 3\theta$ in terms of $\sin\theta$.

19a Express $\tan 2\theta = k$ as a quadratic equation in t, where $t = \tan\theta$.

   b Use the discriminant to show that the roots are real for k real.

Cumulative Revision Exercises

Exercise 1

1   Without using tables find exact values of:

a $\sin x°$ and $\tan x°$, if $\cos x° = \frac{3}{5}$ and $0 < x < 90$.
b $\sin a°$ and $\cos a°$, if $\tan a° = -\frac{12}{5}$ and $0 < a < 180$.

2   Express the following in terms of $\sin \theta$ or $\cos \theta$:

a $\sin(-\theta)$ b $\cos(-\theta)$ c $\sin(\pi - \theta)$

d $\cos(\pi + \theta)$ e $\sin(\frac{1}{2}\pi - \theta)$ f $\cos(2\pi - \theta)$

3   By solving a trigonometrical equation, show that the curves $y = \sin x°$ and $y = \tan x°$ intersect at only one point in the interval $0 \leqslant x \leqslant 90$.

4   If $\sin \alpha = \frac{4}{5}$ and $\tan \beta = \frac{5}{12}$, and α and β are both acute angles, find, without using tables, the values of $\sin(\alpha + \beta)$ and $\tan(\alpha - \beta)$.

5   Show that $\sin(A - B) + \cos(A + B) = (\sin A + \cos A)(\cos B - \sin B)$.

6   Solve for $0 \leqslant x < 360$, $2\cos 2x° + \cos x° - 1 = 0$.

7   Express as a sum or difference, and simplify:

a $2\sin\frac{1}{2}(\alpha + \beta)\cos\frac{1}{2}(\alpha - \beta)$ b $2\cos(x + 105)°\cos(x + 75)°$

8   Express as products, and simplify: a $\cos 3\theta - \cos 7\theta$

b $\sin(x + 2k) - \sin x$ c $\sin 50° + \sin 70°$

9   Solve for $0 \leqslant x \leqslant 360$: $\sin x° + \sin 3x° = 0$.

10a   Express $x = \sin 3\theta + \sin \theta$ and $y = \cos 3\theta + \cos \theta$ as products.
b   Hence show that: (1) $x/y = \tan 2\theta$ (2) $x^2 + y^2 = 2(1 + \cos 2\theta)$

11   Write down the maximum and minimum values of:

a $2\sin x°$ b $\cos^2 x°$ c $2\sin x°\cos x°$ d $2 - 3\cos 5x°$

12a   Express $3\sin x° - 4\cos x°$ as $k\cos(x - \alpha)°$; $k > 0$ and $0 < \alpha < 360$.
b   Hence find the maximum and minimum values of the expression, and the corresponding replacements for x ($0 \leqslant x \leqslant 360$).

13   Solve for $0 \leqslant x < 360$: $8\sin x° + 15\cos x° = 17$.

Exercise 2

1   List the following in increasing order of magnitude:
$$\sin 50°, \sin 70°, \sin \tfrac{1}{3}\pi, \sin \tfrac{1}{4}\pi, \sin 30°, \sin \tfrac{1}{2}\pi.$$

2   P is the point $(a\cos\theta, b\sin\theta)$ and Q is $(b\cos\theta, a\sin\theta)$, where a and b are positive, and $a > b$. Show that the length of PQ is $a-b$.

3   If $\sin\alpha = \tfrac{4}{5}$ and $\cos\beta = \tfrac{12}{13}$, where α is obtuse and β is acute, evaluate, without using tables, $\sin 2\alpha$ and $\cos(\alpha-\beta)$.

4   Given $\tan(A+B) = -7$ and $\tan A = 3$, find $\tan B$.

5   Solve for $0 \leqslant x \leqslant 360$: $\cos 2x° + \sin x° + 2 = 0$.

6   Express as a sum or difference and simplify:

  a $2\cos(\pi+4\theta)\cos(\pi-2\theta)$ *b* $\cos 65° \sin 25°$

7 *a*   Simplify: $E = \dfrac{\sin 5x° + \sin x°}{\cos 5x° + \cos x°}$ *b*   Hence solve $E = -1, 0 < x < 360$.

8   Express $\sin 3\theta - \sin\theta$ as a product, and hence solve the equation $\sin 3\theta - \sin\theta = 2\cos^2 2\theta$, for $0 \leqslant \theta \leqslant 2\pi$.

9   If $p = a\cos 2\theta$ and $q = b\sin\theta$, prove that $b^2 p = a(b^2 - 2q^2)$.

10*a*   Express $\sqrt{3}\sin x° + \cos x°$ in the form $k\cos(x-\alpha)°$, $k > 0$ and $0 < \alpha < 360$.

  b   Hence find the maximum and minimum values of the expression, and the corresponding replacements for x ($0 \leqslant x \leqslant 360$).

11   Solve, for $0 \leqslant x \leqslant 360$:

  a $24\cos x° - 7\sin x° = 20$ *b* $2\sin x° - 3\cos x° = 1$

12   With respect to the usual origin and axes, points A and C have Cartesian coordinates (a, b) and (c, d), and polar coordinates (r, α) and $(r, \alpha+\theta)$ respectively.

  a   Show that $c = a\cos\theta - b\sin\theta$, and express d in terms of a, b and θ.

  b   Find matrix M such that $\begin{pmatrix} c \\ d \end{pmatrix} = M\begin{pmatrix} a \\ b \end{pmatrix}$, and show that $\begin{pmatrix} 0 & -1 \\ 1 & 0 \end{pmatrix}$ is a special case of M. What transformations are effected by the two matrices?

13   Prove that: *a* $\cos\theta - \sin 2\theta \sin\theta = \cos\theta \cos 2\theta$, for all θ.

               b $\sin\theta = \dfrac{2t}{1+t^2}$ and $\cos\theta = \dfrac{1-t^2}{1+t^2}$, where $t = \tan\tfrac{1}{2}\theta$.

Calculus

Further Differentiation and Integration

1 The derivatives of sin x and cos x

We first prove that $\lim\limits_{\theta \to 0} \dfrac{\sin \theta}{\theta} = 1$.

In evaluating the limit we use a geometrical argument in which θ is interpreted as the measure in radians of an angle. Since we are concerned with θ tending to zero, we can suppose that θ is a small positive or negative number and in particular that $-\frac{1}{2}\pi < \theta < \frac{1}{2}\pi$.

The required limit is evaluated by sandwiching $\dfrac{\sin \theta}{\theta}$ between bounds which clearly lead to definite limits as θ approaches zero.

Suppose in the first place that $0 < \theta < \frac{1}{2}\pi$; take a positive angle AOB whose measure in radians is θ.

In Figure 1, OA has length 1 unit and arc AB is part of the circle with centre O and radius 1 unit. BD is the altitude from B and AE is the tangent to the arc at A.

The length of BD $= \sin \theta$ units, and the length of AE $= \tan \theta$ units.

Then $\dfrac{\text{area of sector OAB}}{\text{area of circle}} = \dfrac{\theta}{2\pi}$

$\Rightarrow$ area of sector OAB $= \dfrac{\theta}{2\pi} . \pi . 1^2 = \frac{1}{2}\theta.$

From the diagram it can be seen that

area of triangle OAB $<$ area of sector OAB $<$ area of triangle OAE

$\Leftrightarrow \quad \frac{1}{2}\text{OA} . \text{DB} < \frac{1}{2}\theta < \frac{1}{2}\text{OA} . \text{AE}$

$\Leftrightarrow \quad \frac{1}{2}\sin \theta \quad\;\; < \frac{1}{2}\theta < \frac{1}{2}\tan \theta$

$\Leftrightarrow \qquad\quad 1 < \dfrac{\theta}{\sin \theta} < \dfrac{1}{\cos \theta}$

$\Leftrightarrow \qquad\quad 1 > \dfrac{\sin \theta}{\theta} > \cos \theta.$

The same result can be obtained for $-\frac{1}{2}\pi < \theta < 0$.

$$\text{As } \theta \to 0, \cos \theta \to 1.$$

It follows that $\lim\limits_{\theta \to 0} \dfrac{\sin \theta}{\theta} = 1$.

Notice that in this proof it is essential that θ be the measure *in radians* of an angle. When the use of limiting processes is involved, therefore, as in calculus, measures of all angles *must be in radians*. As it happens, this also turns out to be more convenient in most technological applications.

Now that we know $\lim\limits_{\theta \to 0} \dfrac{\sin \theta}{\theta} = 1$, we are in a position to find the derivatives of the sine and cosine functions. But first let us be quite clear what these functions are. In calculus we are concerned with functions whose domains are the set R of real numbers or subsets of R.

The sine function is defined as $\sin : x \to \sin x$, $x \in R$, where $\sin x = \sin(x \text{ radians})$. In other words, $\sin x$ is the value of the familiar sine function, domain the set of angles, providing x is thought of as the measure *in radians* of an angle. For example, under this mapping the image of $\frac{1}{6}\pi$ is $\sin\frac{1}{6}\pi$, i.e. $\frac{1}{2}$.

The cosine function is defined similarly.

To find the derivatives of the functions defined by:

(i) $f(x) = \sin x$ (ii) $f(x) = \cos x$

Reminders

$$\sin C - \sin D \qquad\qquad \cos C - \cos D$$
$$= 2 \cos \tfrac{1}{2}(C+D) \sin \tfrac{1}{2}(C-D) \qquad = -2 \sin \tfrac{1}{2}(C+D) \sin \tfrac{1}{2}(C-D)$$

(i) $\dfrac{f(x+h) - f(x)}{h}$ (ii) $\dfrac{f(x+h) - f(x)}{h}$

$$= \frac{\sin(x+h) - \sin x}{h} \qquad\qquad = \frac{\cos(x+h) - \cos x}{h}$$

$$= \frac{2 \cos(x+\frac{1}{2}h) \sin \frac{1}{2}h}{h} \qquad\qquad = \frac{-2 \sin(x+\frac{1}{2}h) \sin \frac{1}{2}h}{h}$$

$$= \cos(x+\tfrac{1}{2}h) \left(\frac{\sin \frac{1}{2}h}{\frac{1}{2}h} \right) \qquad\qquad = -\sin(x+\tfrac{1}{2}h) \left(\frac{\sin \frac{1}{2}h}{\frac{1}{2}h} \right)$$

So $f'(x) = \lim_{h \to 0} \dfrac{f(x+h)-f(x)}{h}$

$= \lim_{h \to 0} \left[\cos(x+\tfrac{1}{2}h) \left(\dfrac{\sin \tfrac{1}{2}h}{\tfrac{1}{2}h} \right) \right]$

$= \lim_{h \to 0} \cos(x+\tfrac{1}{2}h) \lim_{h \to 0} \left(\dfrac{\sin \tfrac{1}{2}h}{\tfrac{1}{2}h} \right)$

$= \cos x . 1$

$= \cos x$

So $f'(x) = \lim_{h \to 0} \dfrac{f(x+h)-f(x)}{h}$

$= \lim_{h \to 0} \left[-\sin(x+\tfrac{1}{2}h) \left(\dfrac{\sin \tfrac{1}{2}h}{\tfrac{1}{2}h} \right) \right]$

$= \lim_{h \to 0} \left[-\sin(x+\tfrac{1}{2}h) \right] \lim_{h \to 0} \left(\dfrac{\sin \tfrac{1}{2}h}{\tfrac{1}{2}h} \right)$

$= -\sin x . 1$

$= -\sin x$

If $f(x) = \sin x, f'(x) = \cos x$.   If $f(x) = \cos x, f'(x) = -\sin x$.

In Leibniz notation, $\dfrac{d}{dx}(\sin x) = \cos x$, and $\dfrac{d}{dx}(\cos x) = -\sin x$.

Exercise 1

Differentiate:

1   a $\sin x$ b $4 \sin x$ c $4 + \sin x$ d $x - 2 \sin x$

2   a $\cos x$ b $3 \cos x$ c $5 + \cos x$ d $x^2 - \cos x$

3   a $2 \sin x$ b $5 \cos x$ c $\sin x + \cos x$

  d $x^3 + \sin x$ e $\cos x - x - 1$ f $2 \cos x + 3 \sin x$

4   Write down the values of the following, with the aid of Figure 2.

  a $\sin \tfrac{1}{2}\pi$ b $\sin \pi$ c $\cos 0$ d $\cos \pi$ e $\cos \tfrac{1}{3}\pi$

  f $\sin \tfrac{1}{4}\pi$ g $\sin \tfrac{1}{6}\pi$ h $\cos \tfrac{3}{2}\pi$ i $\sin \tfrac{2}{3}\pi$ j $\sin \tfrac{4}{3}\pi$

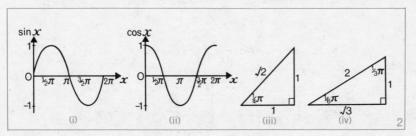

(i)   (ii)   (iii)   (iv)

2

5   Find the gradient of the graph of the function f defined by $f(x) = \sin x + \cos x$ at:

   a $x = 0$ b $x = \frac{1}{4}\pi$ c $x = \frac{1}{2}\pi$ d $x = \pi$

6   Show that the following functions are never decreasing:

   a $f : x \to x + \sin x$ b $g : x \to x + \cos x$

7   Find the equation of the tangent to the curve:

   a $y = \sin x$ at $x = \frac{1}{4}\pi$ b $y = \cos x$ at $x = \frac{1}{6}\pi$

8   Use differentiation to show that the sine function has a maximum turning value 1 at $x = \frac{1}{2}\pi$, and a minimum turning value -1 at $x = \frac{3}{2}\pi$.

9   Show that the cosine function has a maximum turning value 1 at 0 and 2π, and a minimum turning value -1 at π.

10 A function f is given by $f(x) = a \sin x + b \cos x$, where a and b are constants.

   a If f has a stationary value at $x = \frac{1}{4}\pi$, show that $a = b$.
   b Find the rate of change of f at $x = \frac{1}{2}\pi$, given $a = b = -1$.

11 The equation of a curve is $y = \sin x + \cos x$. For $0 \leqslant x \leqslant 2\pi$, find:

   a the points of intersection of the curve and the x-axis.
   b the turning points of the curve, and the nature of each. Hence sketch the curve.

---

2   The integrals of sin x and cos x

If $f(x) = \sin x$, then $f'(x) = \cos x$.
Hence, if $f'(x) = \cos x$, then $f(x) = \sin x + C$.

Using the usual indefinite integral notation for the anti-derivative,

$$\int \cos x \, dx = \sin x + C.$$

In the same way, since $\dfrac{d}{dx}(\cos x) = -\sin x$,

$$\int \sin x \, dx = -\cos x + C.$$

Example. $\displaystyle \int_0^{\pi/2} \cos x \, dx = \Big[\sin x\Big]_0^{\pi/2} = 1 - 0 = 1.$

Exercise 2

Integrate:

1 *a* $\cos x$ *b* $2\cos x$ *c* $1-\cos x$ *d* $x+3\cos x$

2 *a* $\sin x$ *b* $4\sin x$ *c* $4-\sin x$ *d* $2\sin x+3x^2$

3   Find: *a* $\displaystyle\int(\cos x+\sin x)\,dx$ *b* $\displaystyle\int(3\cos x-4\sin x)\,dx$

4   Evaluate:

   a $\displaystyle\int_0^{\pi/6}\cos x\,dx$ *b* $\displaystyle\int_0^{\pi/2}\sin x\,dx$ *c* $\displaystyle\int_0^{\pi}(\cos x+\sin x)\,dx$

5   Evaluate:

   a $\displaystyle\int_0^{\pi/2}(1+\cos x)\,dx$ *b* $\displaystyle\int_0^{\pi/3}(1+2\sin x)\,dx$ *c* $\displaystyle\int_{-\pi}^{\pi}(2x+\sin x)\,dx$

6   Evaluate $\displaystyle\int_0^{\pi}\cos x\,dx$. Show the area given by this integral in a sketch.

7   Evaluate $\displaystyle\int_0^{\pi}\sin x\,dx$. Show the area given by this integral in a sketch.

8   Find the area of the region bounded by the curve $y=1+\sin x$, the x-axis and the lines $x=\frac{1}{3}\pi$ and $x=\frac{1}{2}\pi$.

9   Find the area of the region in the first quadrant enclosed by the y-axis and the curves $y=\sin x$ and $y=\cos x$.

10   Use the formula $\cos 2\alpha=2\cos^2\alpha-1$ to show that $\cos^2\frac{1}{2}x=\frac{1}{2}(1+\cos x)$. Hence find $\displaystyle\int\cos^2\frac{1}{2}x\,dx$.

11   Express $\sin^2\frac{1}{2}x$ in terms of $\cos x$, and hence find $\displaystyle\int\sin^2\frac{1}{2}x\,dx$.

12   Evaluate: *a* $\displaystyle\int_{\pi/2}^{\pi}(x^2+5\sin x)\,dx$ *b* $\displaystyle\int_{-\pi}^{\pi/2}(2\sin x-3\cos x)\,dx$

13   Verify that the curve $y=\sin x$ and the line $y=2x/\pi$ intersect at $x=-\frac{1}{2}\pi$, $x=0$ and $x=\frac{1}{2}\pi$, and sketch the curve and line for $\{x:-\pi\leqslant x\leqslant\pi\}$.

Find the area of the region in the first quadrant enclosed by the curve and the line. What is the total area enclosed by the curve and the line?

Find also the area of the region bounded by the curve, the line and the lines $x = \frac{1}{2}\pi$ and $x = \pi$.

3　The chain rule for differentiating composite functions

Composite functions

At present we can differentiate a function only if it can be given by a formula involving a constant multiple of a power of x, and a sine or a cosine, or a sum of such terms, e.g.

$$f(x) = 2 + 2x^{-1} + 4x^{1/2} + 5\sin x - 6\cos x.$$

Thus to find $\dfrac{d}{dx}(1+x^2)^3$ we should have to expand $(1+x^2)^3$ to give the sum of terms $1 + 3x^2 + 3x^4 + x^6$ and then differentiate term by term. We cannot yet find $\dfrac{d}{dx}(1+x^2)^{1/2}$, and it is the purpose of this section to develop a technique capable of handling such a situation. This technique is closely connected with the composition of functions which we studied in Book 8, so we begin by revising this work.

Example 1. If $F(x) = (1+x^2)^{1/2}$, express F as the composition of two functions f and g, in the form $F(x) = f(g(x))$, defining f and g.

From Figure 3, F is equivalent to the mapping $g: a \to 1 + a^2$ followed by the mapping $f: 1 + a^2 \to (1 + a^2)^{1/2}$. That is,

$$F = f \circ g, \text{ or } F(x) = f(g(x)), \text{ where } g(x) = 1 + x^2 \text{ and } f(x) = x^{1/2}.$$

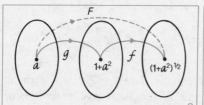

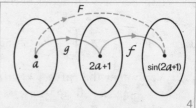

Example 2. If $F(x) = \sin(2x+1)$, write F as the composition of two functions f and g.

From Figure 4, $F(a) = f(g(a)) = f(2a+1) = \sin(2a+1)$.

Hence $F = f \circ g$, or $F(x) = f(g(x))$, where $g(x) = 2x+1$ and $f(x) = \sin x$.

Exercise 3

With the aid of mapping diagrams as in Figures 3 and 4 express $F(x)$ in the form $f(g(x))$, i.e. as the composition of two functions which you can already differentiate, and define $f(x)$ and $g(x)$ in each case.

1   $F(x) = (x+1)^6$ 2 $F(x) = (2x+3)^4$ 3 $F(x) = \sqrt{(x+2)}$

4   $F(x) = \sin(3x+4)$ 5 $F(x) = \cos(ax+b)$ 6 $F(x) = (\sin x)^2$

7   $F(x) = (x^2+x+1)^3$ 8 $F(x) = \sqrt[3]{(x^2-1)}$ 9 $F(x) = (1+\cos x)^3$

10 $F(x) = \dfrac{1}{x+3}$ 11 $F(x) = \left(x+\dfrac{1}{x}\right)^6$ 12 $F(x) = \dfrac{1}{5x-4}$

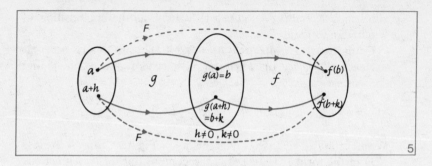

5

The chain rule

Example. If $F(x) = (1+x^2)^{1/2}$, find $F'(x)$.

We can write $F(x) = f(g(x))$, where $g(x) = 1+x^2$ and $f(x) = x^{1/2}$.

$$F'(a) = \lim_{h \to 0} \frac{F(a+h)-F(a)}{h}, \text{ by definition.}$$

$$= \lim_{h \to 0} \frac{f(b+k)-f(b)}{h}, \text{ from Figure 5.}$$

$$= \lim_{h \to 0} \frac{f(b+k)-f(b)}{k} \cdot \frac{k}{h}$$

$$= \lim_{h \to 0} \frac{f(b+k)-f(b)}{k} \cdot \frac{g(a+h)-g(a)}{h}, \text{ since } g(a+h) = b+k$$
$$= g(a)+k$$

$$= \lim_{k \to 0} \frac{f(b+k)-f(b)}{k} \cdot \lim_{h \to 0} \frac{g(a+h)-g(a)}{h}$$

$$= f'(b) \cdot g'(a)$$

Now $f(b) = b^{1/2} \Rightarrow f'(b) = \tfrac{1}{2}b^{-1/2}$
and $g(a) = 1+a^2 \Rightarrow g'(a) = 2a$.

So $F'(a) = \tfrac{1}{2}b^{-1/2} \cdot 2a = \dfrac{a}{(1+a^2)^{1/2}}$, since $b = g(a) = 1+a^2$.

As this is true for each element a in the domain of g (and each element b in the domain of f), we have

$$F'(x) = \frac{x}{(1+x^2)^{1/2}}.$$

It can be shown in the same way that:

If $F = f \circ g$, so that $F(x) = f(g(x))$, where f and g are differentiable functions, then F is differentiable, and $F'(x) = f'(g(x)) \cdot g'(x)$.

This is called the *chain rule* for differentiation since it can be extended to a 'chain' of functions $f \circ g \circ h \circ \ldots$

To differentiate F using this rule it is helpful to think of a peeling process, as with an onion, where we start with the outer 'skin' function and proceed inwards, step by step?

Example 1. Given $F(x) = (x^2 + 1)^6$, find $F'(x)$.

Let $\quad F(x) = f(g(x))$, where $g(x) = x^2 + 1$ and $f(x) = x^6$.
Then $F'(x) = f'(g(x)) . g'(x)$
$$= 6(\underbrace{\ldots}_{g(x)})^5 \times 2x$$
$$= 12x(x^2 + 1)^5.$$

Example 2. Given $\quad F(x) = \left(x - \dfrac{1}{x}\right)^3$, find $F'(x)$.

$$F'(x) = 3\left(x - \frac{1}{x}\right)^2 \times \left(1 + \frac{1}{x^2}\right).$$

In practice the following pattern may be useful:

$$\frac{d}{dx}f(\ldots) = f'(\ldots) \times \frac{d}{dx}(\ldots) \qquad . \qquad . \qquad . \qquad (1)$$

where the same differentiable function is inserted in each bracket.

Example 3. $\dfrac{d}{dx}(3x - 4)^{1/3} = \tfrac{1}{3}(3x - 4)^{-2/3} \times 3 = (3x - 4)^{-2/3}.$

Example 4. $\dfrac{d}{dx}\left(\dfrac{1}{x^2 + 2x + 3}\right) = \dfrac{d}{dx}(x^2 + 2x + 3)^{-1}$
$$= -(x^2 + 2x + 3)^{-2} \times (2x + 2)$$
$$= \frac{-2(x + 1)}{(x^2 + 2x + 3)^2}$$

Note (i) The link between the Leibniz and the 'dash' notation, namely

$$\frac{d}{dx}f(x) = f'(x),$$

arises as the special case of (1) above in which x itself is inserted in each bracket, since $\dfrac{d}{dx}(x) = 1$.

(ii) In Leibniz notation, putting $y = F(x) = f(g(x)) = f(u)$, where $u = g(x)$, $F'(x)$ becomes $\dfrac{dy}{dx}$, $f'(u)$ becomes $\dfrac{dy}{du}$ and $g'(x)$ becomes $\dfrac{du}{dx}$.

The chain rule then appears in the form $\dfrac{dy}{dx} = \dfrac{dy}{du} \cdot \dfrac{du}{dx}$.

Exercise 4

Use the chain rule to differentiate the following:

1 a $(x+1)^6$ *b* $(2x+3)^4$ *c* $(5+x)^3$ *d* $(3x-1)^8$

2 a $(1-x)^4$ *b* $(3-4x)^3$ *c* $(5-x)^7$ *d* $(2-4x)^5$

3 a $(x^2+1)^5$ *b* $(x^3-1)^4$ *c* $(x^4+8)^3$ *d* $(1-x^5)^2$

4 a $(2x+3)^{1/2}$ *b* $(3x-1)^{1/3}$ *c* $(x-1)^{-4}$ *d* $(2-3x)^{-2}$

5 a $(x^2+3x)^3$ *b* $(x^2+x+5)^2$ *c* $(x^3-x)^{10}$ *d* $(x^2+2x+4)^{1/2}$

6 a $\sqrt{(2x-5)}$ *b* $\sqrt{(1-4x)}$ *c* $\sqrt[3]{(x^3+3x)}$ *d* $\sqrt[4]{(2-x^2)}$

7 a $\dfrac{1}{x+1}$ *b* $\dfrac{5}{1-x}$ *c* $\dfrac{1}{(2x+3)^2}$ *d* $\dfrac{4}{\sqrt{(4x+1)}}$

8 a $(x^2+2)^2$ *b* $\left(x+1+\dfrac{1}{x}\right)^2$ *c* $\left(x^2+1+\dfrac{1}{x^2}\right)^4$

9   Find the stationary value of the function f defined by $f(x) = 8x - (2x-1)^{4/3}$, and determine its nature.

10   Show that if $f(x) = (x-1)^{3/2} - 3(x-1)^{1/2}$, then $f'(x) = \dfrac{3(x-2)}{2(x-1)^{1/2}}$.

Find the stationary value of the function, and determine its nature.

*   *   *   *

Example 5. Differentiate: (i) $\sin(2x+1)$ (ii) $\cos^3 x$.

The derivatives are:

(i) $\cos(2x+1) \times 2$
   $= 2\cos(2x+1)$

(ii) $3\cos^2 x \times (-\sin x)$
   $= -3\cos^2 x \sin x$

Exercise 5

Differentiate the following (questions *1–4*):

1 a $\sin 3x$ **b** $\sin(4x+1)$ **c** $\sin(ax+b)$ **d** $\sin\frac{1}{2}x$

2 a $\cos 2x$ **b** $\cos(3x+4)$ **c** $\cos(px+q)$ **d** $\cos\frac{1}{2}x$

3 a $\sin^2 x$ **b** $\sin^3 x$ **c** $\cos^2 x$ **d** $\cos^4 x$

4 a $(1+\sin x)^2$ **b** $(1-\cos x)^3$ **c** $\dfrac{2}{\sin x}$ **d** $\sin^2 x + \cos^2 x$

5   Which of the following are true and which are false?

   a $\dfrac{d}{dx}\sin 2x = \cos 2x$ **b** $\dfrac{d}{dx}\cos(x-1) = -\sin(x-1)$

   c $\dfrac{d}{dx}(\sin^2 x - \cos^2 x) = 2\sin 2x$

   d $\dfrac{d}{dx}(\sin mx + \cos nx) = m\sin x - n\cos x$

6   Show that the function defined by $f(x) = x + 2\cos\frac{1}{2}x$ is never decreasing. Find x for stationary values of f in the interval $0 \leqslant x \leqslant 2\pi$.

7   Find x for stationary values of these functions for $0 \leqslant x \leqslant 2\pi$:

   a $f(x) = 4\sin x + \cos 2x$ **b** $f(x) = 2\sin x + \sin 2x$

8   Find the points of intersection with the axes, and the turning points, of the curve $y = \sin 2x$ in the interval $0 \leqslant x \leqslant \pi$. Hence sketch the curve.

9   Repeat question **8** for the curve $y = \cos 2x$.

10a If $F(x) = \sin x^\circ$, find $F'(x)$. (*Hint.* Construct $F = f \circ g$, where $f : x \to \sin x$ and $g : x \to \dfrac{\pi}{180}x$.)

   b Repeat **a** for $F(x) = \cos x^\circ$.

4   Some special integrals

In the theory of integration there is *no* formula resembling the chain rule. Integration is in general less straightforward than differentiation, and it is advisable to check an integration by differentiating back.

It may be that in the course of differentiating back only one application of the chain rule will be needed, and furthermore that this will produce only *a numerical factor*, e.g. $\dfrac{d}{dx}(\sin 3x) = \cos 3x \,.\, 3$. In such *very special* cases we are already able to perform the integration: thus $\int \cos 3x \, dx = \tfrac{1}{3} \sin 3x + C$ as may be checked by differentiating.

Example 1. $\dfrac{d}{dx}(2x+1)^3 = 3(2x+1)^2 \times 2 = 6(2x+1)^2$

Hence $\int (2x+1)^2 dx = \tfrac{1}{6}(2x+1)^3 + C$

In general, $\int (ax+b)^n \, dx = \dfrac{1}{a(n+1)}(ax+b)^{n+1} + C,\ a \neq 0,\ n \neq -1.$

Proof. $\dfrac{d}{dx}\left(\dfrac{1}{a(n+1)}(ax+b)^{n+1}\right) = \dfrac{1}{a(n+1)}(n+1)(ax+b)^n \times a$

Example 2. Evaluate $\displaystyle\int_0^2 \dfrac{dx}{\sqrt{(4x+1)}}$

$$\int_0^2 (4x+1)^{-1/2}\, dx = \left[\dfrac{1}{4 \times \frac{1}{2}}(4x+1)^{1/2}\right]_0^2 = \tfrac{1}{2}(3-1) = 1$$

Example 3. $\dfrac{d}{dx}(x^2+1)^2 = 2(x^2+1)\,.\,2x = 4x(x^2+1)$

But $\int (x^2+1)\, dx \neq \dfrac{1}{4x}(x^2+1)^2 + C$ as can be seen by multiplying out the right-hand side and differentiating term by term. In this case notice that one application of the chain rule is sufficient, but $4x$ is not simply 'a numerical factor'.

Exercise 6

Integrate:

1 a $(x+3)^4$ *b* $(2x+1)^6$ *c* $(ax+b)^4$ *d* $(1-2x)^3$

2 a $(x+4)^{-2}$ *b* $(3x+2)^{-4}$ *c* $(2x-1)^{1/2}$ *d* $(1-2x)^{3/2}$

3 a $\dfrac{1}{(x+1)^2}$ *b* $\dfrac{2}{(x-3)^3}$ *c* $\dfrac{1}{\sqrt{(2x+5)}}$ *d* $\dfrac{1}{\sqrt[3]{(x-1)^2}}$

Evaluate:

4 a $\displaystyle\int_0^1 (2x+1)^3 \, dx$ *b* $\displaystyle\int_{-1}^1 (1-x)^3 \, dx$ *c* $\displaystyle\int_0^4 \sqrt{(4-x)} \, dx$

5 a $\displaystyle\int_1^2 \dfrac{dx}{(x+2)^2}$ *b* $\displaystyle\int_{-1}^3 \dfrac{dx}{\sqrt{(2x+3)}}$ *c* $\displaystyle\int_{-1}^4 (3x+4)^{3/2} \, dx$

Example 4. $\dfrac{d}{dx}\sin(2x+1) = 2\cos(2x+1)$

Hence $\int \cos(2x+1)\,dx = \tfrac{1}{2}\sin(2x+1)+C$

In general, $\int \cos(ax+b)\,dx = \dfrac{1}{a}\sin(ax+b)+C,\ a \neq 0$

$$\int \sin(ax+b)\,dx = -\dfrac{1}{a}\cos(ax+b)+C, \quad a \neq 0.$$

Example 5. $\int \sin 3x \cos x\, dx = \tfrac{1}{2}\int (\sin 4x + \sin 2x)\,dx$

$= \tfrac{1}{2}(-\tfrac{1}{4}\cos 4x - \tfrac{1}{2}\cos 2x)+C = -\tfrac{1}{8}\cos 4x - \tfrac{1}{4}\cos 2x +C.$

Exercise 7

Integrate:

1 a $\cos 2x$ *b* $\cos 4x$ *c* $\cos(x+5)$ *d* $\cos(3x+2)$

2 a $\sin 3x$ *b* $-\sin 5x$ *c* $\sin\tfrac{1}{2}x$ *d* $\sin(2-x)$

Evaluate:

3 a $\displaystyle\int_0^{\pi/6} \cos 3x\, dx$ *b* $\displaystyle\int_0^{\pi/4} \sin 2x\, dx$ *c* $\displaystyle\int_{-\pi/4}^{\pi/4} \cos(3x+\tfrac{1}{4}\pi)\,dx$

4 a $\displaystyle\int_0^{\pi/2} (\cos 2x - \sin 2x)\,dx$ **b** $\displaystyle\int_0^{\pi/12} (\cos 6x + \sin 4x)\,dx$

5 Evaluate $\displaystyle\int_0^{\pi/2} \cos 2x\,dx$. Show the area given by this integral in a sketch.

6 Evaluate $\displaystyle\int_0^{\pi/2} \sin 2x\,dx$. Show the area given by this integral in a sketch.

7 a Express each of the following in terms of $\sin 2x$ or $\cos 2x$:

    (1) $\sin x \cos x$ (2) $\cos^2 x$ (3) $\sin^2 x$ (4) $\cos^2 x - \sin^2 x$

  b Hence find: (1) $\int \sin x \cos x\,dx$ (2) $\int \cos^2 x\,dx$ (3) $\int \sin^2 x\,dx$

    (4) $\int (\cos^2 x - \sin^2 x)\,dx$

8 Express $\cos 3x \cos x$ as a sum of two cosines. Hence find $\int \cos 3x \cos x\,dx$

9 Find: *a* $\int \sin 4x \cos 2x\,dx$ *b* $\int \cos 2x \cos x\,dx$ *c* $\int \sin 5x \sin x\,dx$

10 Find the volume of the solid generated when the region enclosed by the curve $y = \sin x$ and the x-axis between $x = 0$ and $x = \pi$ is rotated through a complete revolution about the x-axis.

11 The area in the first quadrant which is enclosed by the curve $y = \cos x$ and the line $y = 1 - \dfrac{2}{\pi}x$ is rotated through one revolution about the x-axis. Find the volume of the solid generated.

12 The derivative at x of function f has been denoted by $f'(x)$, or $\dfrac{df}{dx}$. Starting with the function f' we can form *its* derived function (if it exists), and this is denoted by f''. The value of f'' at x is denoted by $f''(x)$ or $\dfrac{d^2 f}{dx^2}$.

  a If $y = \sin nx$, show that $\dfrac{d^2 y}{dx^2} + n^2 y = 0$.

  b Show that this equation is also satisfied by $y = a \cos nx + b \sin nx$ for every pair of constants a, b.

  c Hence find a solution of the equation for which $y = 2$ and $\dfrac{dy}{dx} = 1$ when $x = 0$.

Summary

1   *The derivatives of the sine and cosine functions*

If $f(x) = \sin x$, then $f'(x) = \cos x$; or $\dfrac{d}{dx}(\sin x) = \cos x$.

If $f(x) = \cos x$, then $f'(x) = -\sin x$; or $\dfrac{d}{dx}(\cos x) = -\sin x$.

2   *The integrals of the sine and cosine functions*

$\int \cos x\, dx = \sin x + C; \quad \int \sin x\, dx = -\cos x + C$

3   *The chain rule for differentiating composite functions*

If $F(x) = f(g(x))$, then $F'(x) = f'(g(x)) \cdot g'(x)$, or

if $y = F(x) = f(g(x))$, and $u = g(x)$, then $\dfrac{dy}{dx} = \dfrac{dy}{du} \cdot \dfrac{du}{dx}$.

4   *Some special integrals*

$\int (ax+b)^n\, dx \quad = \quad \dfrac{1}{a(n+1)}(ax+b)^{n+1} + C,\, a \neq 0,\, n \neq -1$

$\int \cos(ax+b)\, dx = \quad \dfrac{1}{a}\sin(ax+b) + C,\, a \neq 0$

$\int \sin(ax+b)\, dx = -\dfrac{1}{a}\cos(ax+b) + C,\, a \neq 0$

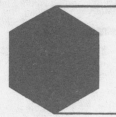

Revision Exercises

Revision Exercise on Chapter 1
Further Differentiation and Integration

Revision Exercise 1

1   Write down the values of:

  a $\sin 0$ *b* $\sin \frac{1}{3}\pi$ *c* $\cos \frac{2}{3}\pi$ *d* $\cos \frac{1}{6}\pi$ *e* $\sin 2\pi$

Differentiate:

2 *a* $10 \sin x$ ***b*** $8 \cos x$ ***c*** $2x - 3 \sin x$ ***d*** $x^2 - \cos x$

3 *a* $3 \cos x - 4 \sin x$ ***b*** $2 \sin x - 2x$ ***c*** $1 - \sin x - \cos x$

4 *a* Find the gradient of the tangent to the curve $y = x + \cos x$ at the points where x has the values: *(1)* 0 *(2)* $\frac{1}{4}\pi$ *(3)* $\frac{1}{2}\pi$ *(4)* $\frac{3}{4}\pi$ *(5)* π.
  b Find the stationary point of the curve in the interval $0 \leqslant x \leqslant \pi$.
  c Show that the function never decreases in this interval, and hence state the nature of the stationary point.

5   The equation of a curve is $y = \sin x - \cos x$. For $0 \leqslant x \leqslant 2\pi$, find:

  a the points of intersection of the curve and the *x*-axis
  b the turning points of the curve, and the nature of each. Sketch the curve.

6   Find: *a* $\int 3 \cos x \, dx$ *b* $\int 7 \sin x \, dx$ *c* $\int (\cos x - \sin x) \, dx$
     d $\int (1 + x + \cos x) \, dx$ *e* $\int (x^2 - 2 \cos x) \, dx$ *f* $\int (6 \sin x - 8 \cos x) \, dx$

7   Evaluate:

a $\displaystyle\int_0^{\pi/3} \sqrt{3}\cos x\,dx$ b $\displaystyle\int_0^{\pi/4} \sqrt{2}\sin x\,dx$ c $\displaystyle\int_0^{\pi/2} (\sin x + \cos x)\,dx$

d $\displaystyle\int_{\pi/3}^{\pi} (x + \cos x)\,dx$ e $\displaystyle\int_{\pi/2}^{\pi} (3\sin x - 4\cos x)\,dx$

8   Find the area of the region bounded by the curve $y = 1 - \cos x$, the x-axis, and the ordinates at $x = \frac{1}{2}\pi$ and $x = \pi$.

9   Verify that the curve $y = 5\sin x$ and the line $y = 3x/\pi$ intersect at $x = 0$ and $x = \frac{5}{6}\pi$. Calculate the area of the region enclosed by the line and the curve for $0 \leqslant x \leqslant \frac{5}{6}\pi$.

10   Differentiate:

a $(3x+4)^3$ b $(2x-1)^7$ c $(1-2x)^{-3}$ d $(x^2+1)^3$

e $(1-3x^2)^9$ f $(2x-1)^{1/2}$ g $\sqrt{(x^3+1)}$ h $(x^2+2x)^{-4}$

11   Find the derivatives of:

a $\sin(2x-1)$ b $3\cos(3x+2)$ c $\sin(x^2)$ d $\frac{1}{2}\cos(x^2+2x)$

e $\sin^4 x$ f $\cos^3 x$ g $(2\sin x+1)^4$ h $(2-3\cos x)^{-1/3}$

12   Integrate: $(4x+3)^5$ $\sqrt{(2x+1)}$ $(2-5x)^{-3}$ $2/(3x+2)^{5/3}$

13   Find:

a $\int \sin(4x+3)\,dx$ b $\int \cos(1-2x)\,dx$ c $\int (\sin 3x + \cos 4x + 5)\,dx$

14   Evaluate:

a $\displaystyle\int_1^5 \sqrt{(2x-1)}\,dx$ b $\displaystyle\int_{-\pi/4}^{\pi/4} 4\cos 2x\,dx$ c $\displaystyle\int_0^{\pi/6} (\sin 3x + \cos 3x)\,dx$

15a Find the area bounded by the curve $y = (3x-2)^3$, the x-axis, and the lines $x = 1$ and $x = 2$.

b   Find the volume of the solid generated by rotating this area through 2π radians about the x-axis.

16   Find the stationary value of the function f defined by

$$f(x) = \frac{1}{2x-1} - \frac{1}{2x+1}, \text{ and determine its nature.}$$

17 If a, b and n are non-zero constants, with $a^2 \neq n^2$ and $b^2 \neq n^2$, show that only two of the replacements: $s = a \cos(nt + b)$, $s = b \cos(at + n)$, $s = n \sin(at + b)$, $s = b \sin(nt + a)$, satisfy the differential equation $\dfrac{d^2 s}{dt^2} + n^2 s = 0$.

18 Figure 1 shows a semicircular swimming pool of radius r metres. A boy can swim steadily at 1 m/s and walk at 3 m/s. He starts at A, swims to P, and then walks round the edge of the pool from P to B. If the angle OAP $= \theta$ radians, show that the time taken in seconds to go from A to P to B is given by $f(\theta) = 2r \cos\theta + \tfrac{2}{3} r\theta$.

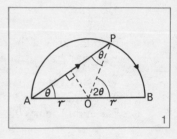

1

Hence find θ for the path which takes the longest time, and for the path which takes the shortest time. Calculate these times for $r = 15$. (Note that $0 \leqslant \theta \leqslant \tfrac{1}{2}\pi$, and remember to consider the end points of the closed interval.)

Cumulative
Revision
Section

Cumulative Revision Section (Books 8-9)

Revision Topic 1. *Differentiation*

Reminders

1 *Rate of change.* The rate of change of the function f at $x = a$ is given by
$$f'(a) = \lim_{h \to 0} \frac{f(a+h) - f(a)}{h}.$$

2 *Derived function, or derivative,* of f is given by
$$f'(x) = \lim_{h \to 0} \frac{f(x+h) - f(x)}{h}.$$

3 *Particular derivatives.* If $f(x) = x^n$, then $f'(x) = nx^{n-1}$, $n \in Q$.

4 *Gradient of tangent to curve* $y = f(x)$ at point (x, y) is $\dfrac{dy}{dx} = f'(x)$.

5 *Increasing and decreasing functions.* If $f'(x) > 0$, f is increasing; if $f'(x) < 0$, f is decreasing.

6 *Stationary value of a function.* If $f'(a) = 0$, then $f(a)$ is a stationary value of f at $x = a$. The *nature* of the stationary value is found by considering changes in the sign of $f'(x)$ in the neighbourhood of $x = a$.

7 *Curve sketching.* Investigate:
   (i) the points where the curve cuts the x- and y-axes
   (ii) stationary points $\left(\text{given by } \dfrac{dy}{dx} = 0 \right)$, and their nature
   (iii) values of y for large positive and negative x.

8 *The derivatives of the sine and cosine functions.*
If $f(x) = \sin x$, then $f'(x) = \cos x$; or $\dfrac{d}{dx}(\sin x) = \cos x$.

If $f(x) = \cos x$, then $f'(x) = -\sin x$; or $\dfrac{d}{dx}(\cos x) = -\sin x$.

9 *The chain rule for differentiating composite functions*
If $F(x) = f(g(x))$, then $F'(x) = f'(g(x)) \cdot g'(x)$, or $\dfrac{dy}{dx} = \dfrac{dy}{du} \cdot \dfrac{du}{dx}$.

Exercise 1

1 Differentiate from first principles the functions f defined as follows:

$$\left[\text{i.e. using the definition } f'(x) = \lim_{h \to 0} \frac{f(x+h)-f(x)}{h}.\right]$$

a $f(x) = 4x$ b $f(x) = 3x^2$ c $f(x) = 2x - 3$

d $f(x) = \dfrac{4}{x}$ e $f(x) = \sin 2x$ f $f(x) = \cos 3x$

2 The distance in metres travelled by a racing car in t seconds is a function of the time, given by $f(t) = 3t^2$. Find the speed of the car at $t = 12$ by evaluating $\lim\limits_{h \to 0} \dfrac{f(12+h)-f(12)}{h}$.

3 Differentiate the following with respect to the appropriate variables:

a $5x^3 + 4x^2 + 3x$ b $t^4 - 2t^2 + 3$ c $\frac{2}{3}u^3 + \frac{3}{2}u^2$

d $4x^{3/2} - 2x^{1/2}$ e $1 - x - \dfrac{1}{x}$ f $\dfrac{1}{x^2} + \dfrac{1}{x^3}$

g $\sqrt{x} - \dfrac{1}{\sqrt{x}}$ h $\dfrac{3}{x} - \dfrac{1}{2x^2}$ i $u^{1/2}(u^{1/2} - u^{-1/2})$

4 Find the derivative of each of the following:

a $(3x+4)^2$ b $(5-4x)^6$ c $\sqrt{(3x-1)}$

d $\sqrt[3]{(6x^2-5)}$ e $x(x+1)^2$ f $x^{1/3}(x^{2/3} + x^{-1/3})$

g $\dfrac{1}{3x-5}$ h $\dfrac{1}{(x+5)^2}$ i $\dfrac{2}{\sqrt{(3-x)}}$

j $\dfrac{x+1}{x}$ k $\dfrac{(x+1)(x+2)}{x}$ l $\dfrac{x^{1/2}+1+x^{-1/2}}{x^{1/2}}$

5 Differentiate the following:

a $\sin 3x$ b $\sin(px+q)$ c $3\sin x - 4\cos x$

d $\sin mx + \cos nx$ e $\cos^2 x$ f $\sin^3 x$

g $\sin^4 x - \cos^4 x$ h $\dfrac{1}{\sin x}$ i $\dfrac{1}{\cos x}$

Exercise 2 (Applications of differentiation)

1   If $f(x) = 3x^2 + 4x - 5$, find $f'(0), f'(-1)$ and $f'(2)$.

2 a   If $f(x) = x + \dfrac{1}{x}$, find $f'(2)$ and $f'(\tfrac{1}{2})$.

   b   Find x for which $f'(x) = 0$, and the corresponding values of $f(x)$.

3   Show that the function $f : x \rightarrow x^3 - 6x^2 + 12x + 5$ is never decreasing.

4   If $f(x) = x^3 - 3x^2 - 9x + 4$, find the stationary points of f, and the range of x for which f is: **a** increasing   **b** decreasing.

5   Find the equations of the tangents to the curves:

   a $y = 4x^2 - 5x + 6$, at $x = -1$ **b** $y = \cos x$, at $x = \tfrac{1}{4}\pi$

   c $y = x^{1/2}(x+3)$, at $x = 4$ **d** $y = \dfrac{1}{x}(3x^2 + 4)$, at $x = 2$

6   A curve has equation $y = ax^2 + b$, where a and b are constants. If the gradient of the tangent to the curve at the point $(3, 4)$ on it is 6, find a and b.

7   Find the maximum and minimum values of $f(x) = x^2 - 2x - 8$ in the closed interval $\{x : 1 \leqslant x \leqslant 5\}$.

8   Find the maximum and minimum values of $f(x) = x^2(6 - x)$ in the closed interval $\{x : 1 \leqslant x \leqslant 5\}$.

9   Show that the function f given by $f(x) = 2x + \sin x$ is never decreasing. Give the gradient of the tangent to its graph at:

   a $x = \tfrac{1}{2}\pi$ **b** $x = \pi$.

10   If $f : x \rightarrow \sin^2 x$ $(0 \leqslant x \leqslant 2\pi)$, find the stationary points of f, and the range of x for which f is: **a** increasing   **b** decreasing. Illustrate with a sketch.

11   Find the stationary values of each of the following functions, and determine the nature of each. Sketch the graphs of the functions.

   a $f(x) = (x-1)^2(x+2)$ **b** $f(x) = x^2(3-x)$

   c $f(x) = x^4 - 8x^2 + 16$ **d** $f(x) = \cos\tfrac{1}{2}x, \ 0 \leqslant x \leqslant 2\pi$.

12   Show that the curve $y = x^3 - 3x^2 + 3x - 9$ has only one point of intersection with the x-axis, and only one stationary point, which is a point of inflexion. Sketch the curve.

13a Show that the curve $y = x^3 - 3x + 1$ cuts the x-axis between $x = -2$ and $x = -1$, between $x = 0$ and $x = 1$, and between $x = 1$ and $x = 2$.

 b Find the stationary points of the curve, and determine the nature of each.

 c Sketch the curve.

14 The height h metres reached by a ball after t seconds is given by the equation $h = 30t - 5t^2$.

 a Find its velocity v metres per second after 3 seconds and 5 seconds; explain your results.

 b Find t when: (1) h is zero (2) v is zero.

 c Calculate the maximum height of the ball.

15 The displacement s centimetres at time t seconds of a particle moving in a straight line is given by the formula $s = t^3 + t^2 - 6$. Calculate:

 a its displacement at $t = 2$ and $t = 3$

 b its velocity and its acceleration at $t = 2$ and $t = 3$

 c its velocity when its acceleration is $5\,\text{cm/s}^2$.

16 The area of a rectangle is $12\,\text{cm}^2$. If one side is x cm long, show that the perimeter P cm is given by $P = 2x + \dfrac{24}{x}$. Find the dimensions of the rectangle which has minimum perimeter.

17 ABCD is a square of side 10 cm. E is a point on BA such that BE = x cm, and F is a point on BC such that CF = $2x$ cm. Show that if the area of triangle DEF is $A\,\text{cm}^2$, then $A = 50 - 10x + x^2$. Find the minimum area of the triangle.

18 A circular sheet of paper is slit along a radius and folded into the shape of a conical container. If the radius of the circle is 6 cm, and the cone has height h cm and volume $V\,\text{cm}^3$, show that $V = \frac{1}{3}\pi h(36 - h^2)$. Find the dimensions of the cone of maximum volume that can be formed in this way.

Revision Topic 2. Integration

Reminders

1 *Anti-derivatives; integrals.*

A function F such that $F'(x) = f(x)$ is called an anti-derivative of f.
The indefinite integral $\int f(x)\,dx = F(x) + C$, C a constant.

The definite integral $\displaystyle\int_a^b f(x)\,dx = \left[F(x)\right]_a^b = F(b) - F(a)$.

2 *Integration.* If $f(x) = x^n$, $\int x^n\,dx = \dfrac{x^{n+1}}{n+1} + C$, $n \neq -1$.

3 *The magnitude of the area of the region* between the curve $y = f(x)$,

the x-axis and the ordinates $x = a$ and $x = b$ is $\displaystyle\int_a^b f(x)\,dx$.

The area between two curves is $\displaystyle\int_a^b [f(x) - g(x)]\,dx \cdot (f(x) \geqslant g(x))$.

4 *The volumes of revolution* through $360°$ of regions are given by:

(i) $\pi \displaystyle\int_a^b y^2\,dx$, about the x-axis (ii) $\pi \displaystyle\int_c^d x^2\,dy$, about the y-axis

(iii) $\pi \displaystyle\int_a^b (y_1{}^2 - y_2{}^2)\,dx$, about the x-axis, for a region between two

curves.

5 *The integrals of the sine and cosine functions.*

$\int \cos x\,dx = \sin x + C$; $\int \sin x\,dx = -\cos x + C$.

6 *Some special integrals*

$\int (ax+b)^n\,dx = \dfrac{1}{a(n+1)}(ax+b)^{n+1} + C$, $a \neq 0$, $n \neq -1$

$\int \cos(ax+b)\,dx = \dfrac{1}{a}\sin(ax+b) + C$, $a \neq 0$

$\int \sin(ax+b)\,dx = -\dfrac{1}{a}\cos(ax+b) + C$, $a \neq 0$

Exercise 3

1   Integrate:

   a $8x^3$ *b* x^2-5 *c* $10x^4+4x$ *d* x^{-2}

   e $4x^3-\dfrac{4}{x^3}$ *f* $\cos 3x$ *g* $\cos x-\sin x$ *h* $\sin(4x-3\pi)$

   i $\cos(\pi-2x)$ *j* $\sin 4x-\cos 4x$ *k* $\cos^2 x$ *l* $\sin^2 x$

2   Find:

   a $\int(5x^4-3x^2+1)\,dx$ *b* $\int(1+2t+3t^2+4t^3)\,dt$ *c* $\int(\tfrac{4}{5}u^3-\tfrac{5}{4}u^4)\,du$

   d $\int(5x^{3/2}-15x^{1/2})\,dx$ *e* $\int(x-3)(2-x)\,dx$ *f* $\int(v+1)(4-3v^2)\,dv$

   g $\int(x^{1/2}-1)^2\,dx$ *h* $\int\left(\dfrac{1}{u^2}-\dfrac{1}{u^3}\right)du$ *i* $\int\left(\sqrt{z}+\dfrac{1}{\sqrt{z}}\right)dz$

3   Find $f(x)$ for each of the following:

   a $f'(x)=4x-1$, and $f(-1)=5$ *b* $f'(x)=x^2+\dfrac{1}{x^2}$, and $f(1)=1$

4   Show by shading in sketches the regions with areas given by:

   a $\displaystyle\int_0^5 x\,dx$ *b* $\displaystyle\int_1^3 x^2\,dx$ *c* $\displaystyle\int_{-2}^2 x^3\,dx$ *d* $\displaystyle\int_0^\pi \sin x\,dx$

5   Find:

   a $\int(2x+3)^4\,dx$ *b* $\int(3u-5)^6\,du$ *c* $\int\sqrt{(5-2t)}\,dt$

   d $\int\dfrac{dx}{\sqrt{(4+x)}}$ *e* $\int\dfrac{dz}{(1-z)^3}$ *f* $\int\cos(3x-\tfrac{1}{4}\pi)\,dx$

   g $\int(x^3-\sin x)\,dx$ *h* $\int(\cos 2x+\sin x)\,dx$ *i* $\int(\cos x-\sin 3x)\,dx$

   j $\int\cos 4x\cos 2x\,dx$ *k* $\int\sin 3x\cos x\,dx$ *l* $\int\sin 2x\sin 6x\,dx$

6   Evaluate the following definite integrals:

   a $\displaystyle\int_0^1(x^2+x+1)\,dx$ *b* $\displaystyle\int_1^2\left(x-\dfrac{1}{x^2}\right)dx$ *c* $\displaystyle\int_{-2}^1\sqrt{(3+t)}\,dt$

   d $\displaystyle\int_0^5\dfrac{dv}{\sqrt{(v+4)}}$ *e* $\displaystyle\int_0^{\pi/2}10\sin 5x\,dx$ *f* $\displaystyle\int_0^{\pi/2}\cos\tfrac{1}{2}t\,dt$

   g $\displaystyle\int_0^{\pi/4}\cos^2 x\,dx$ *h* $\displaystyle\int_0^{\pi/4}\cos 3x\cos x\,dx$ *i* $\displaystyle\int_0^\pi\cos 2x\sin x\,dx$

Exercise 4 (Applications of integration)

In each of questions *1–7*, calculate the total area enclosed by the curve and the line. Illustrate by sketches where necessary.

1   The parabola $y = 3 - 2x - x^2$ and the x-axis.

2   The curve $y = x^2(x + 3)$ and the x-axis.

3   The curve $y = x(x - 1)(x - 3)$ and the x-axis.

4   The curve $y = (x - 1)^2(x - 4)$ and the x-axis.

5   The parabola $y = 3x - x^2$ and the line $y = x - 8$.

6   The curve $y = \sin 2x$ and the x-axis in the interval $\{x : 0 \leqslant x \leqslant \frac{1}{2}\pi\}$.

7   The curve $y = \sin x + \cos x$ and the x-axis in the interval $\{x : 0 \leqslant x \leqslant \frac{1}{2}\pi\}$.

8   The gradient of the tangent to a curve at each point (x, y) on it is given by $\dfrac{dy}{dx} = \sqrt{(2x + 1)}$. If the curve passes through the point $(4, 10)$, find its equation.

9   A particle starts from rest at time $t = 0$ and has an acceleration a metres per second per second at time t seconds, given by the formula $a = \frac{1}{2}t$. Calculate the velocity and displacement at $t = 2$.

10   The region in the first quadrant enclosed by the parabola $y^2 = x$, the x-axis and the line $x = 2$ is rotated through one revolution about the x-axis. Find the volume of the solid formed.

11   Find the volume of the solid revolution formed by rotating the region bounded by the curve $y = \cos x$ and the x-axis, from $x = -\frac{1}{2}\pi$ to $x = \frac{1}{2}\pi$ about the x-axis, through one revolution.

12   If the region enclosed by the parabola $y^2 = x$, the y-axis and the line $y = 1$ is rotated through one revolution about the y-axis, find the volume of the solid formed.

13   Find the area of the region enclosed by the parabola $y^2 = x$ and the line $x = 4$. If this region is rotated through one revolution about the y-axis, find the volume of the solid generated.

14   The region enclosed by the curve $y = \sin 2x$ and the line $y = 4x/\pi$ from $x = 0$ to $x = \frac{1}{4}\pi$ is rotated through one revolution about the x-axis. Find the volume of the solid generated.

Answers

Answers

Page 4   *Exercise 1*

1   $\{1,3\}$ 2 $\{-1,1\frac{1}{2}\}$ 3 $\{0,-1\frac{1}{2}\}$ 4 $\{0,5\}$ 5 $\{-2,2\}$ 6 $\{-\frac{1}{2},\frac{1}{2}\}$

7   $\{-4\}$ 8 $\{-5,3\frac{1}{2}\}$ 9 $\{2,-1\frac{2}{3}\}$ 10 $\{-1,3\}$ 11 $\{-2\cdot41,0\cdot41\}$

12   $\{-1\cdot16,5\cdot16\}$ 13 $\{0\cdot29,1\cdot71\}$ 14 $\left\{\dfrac{-b+\sqrt{(b^2-4ac)}}{2a},\dfrac{-b-\sqrt{(b^2-4ac)}}{2a}\right\}$

15   $\{1,3\}$ 16 $\{-1,1\frac{1}{2}\}$ 17 $\{-2,2\}$ 18 ϕ 19 $\{-0\cdot85,2\cdot35\}$

20   $\{-1\cdot47,0\cdot27\}$

Page 6   *Exercise 2*

1 real, rational	2 not real	3 equal, real, rational
4 real, irrational	5 real, rational	6 not real
7 real, rational	8 real, irrational	9 equal, real, rational
10 not real	11 real, rational	12 equal, real, rational
13 not real	14 real, rational	15 not real
16 real, irrational	17 equal, real, rational	18 not real

19   yes, yes, no, no; yes, yes, no, yes; no, no, no, no; yes, no, yes, no

Page 7   *Exercise 3*

1   a $k=16$ b $k=4$ c $k=\pm8$ 2   a $k\leqslant16$ b $k\leqslant4$ c $k\geqslant8$ or $k\leqslant-8$

3   a $m=\pm3$ b $-3<m<3$ 4 $p=-7$ or 5 5 $-1<q<1$

6   $p=-1\frac{1}{2}$ or 3 7 $a=1$ or $-\frac{1}{3}$; $\{-2\},\{-6\}$ 8 $pq=2$ 9 $c=\pm4$

10   Discriminant $=1+4c^2>0$ for all real c.   12a $m=1$ or 5 b   not real

13   $m=\pm1\frac{1}{2}$ 15 $x^2+(4-2n)x+(10-5n)=0$ 17 $k=9,16,21,24,25$

Page 10   *Exercise 4*

1   $y=2x-1$ 2 $y=-8x-8$ 3 $y=-4x+4$ 4 $y=2x$

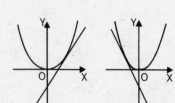
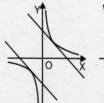
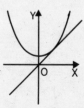

5 $y = x + 1$ *6* $y = -x \pm 4$ *7* $y = \pm 4x - 4$ *8* $y = \pm 8x - 2$

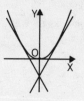
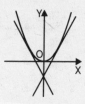

9 $y = \pm 4x$ *10* $y = -x + 2$ *11* $y = \pm \sqrt{3}x - 4$ *12* $y = \pm \frac{4}{3}x + 5$

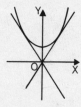
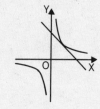
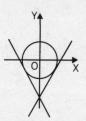
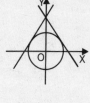

13 $k = 1; (-2, 1)$ *14a* $k = -1; (3, -1)$ *b* $k = 9; (4, 9)$ *15* $c = -4; (2, -2)$

Page 12   *Exercise 5*

1 a (ii)   *b* (iv)   *c* (i)   *d* (v)   *e* (iii)   *f* (vi)

2   *3*   *4*   *5*   *6*

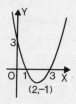

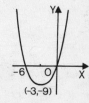

7   *8*   *9*   *10*   *11*

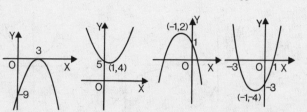

12   13   14   15   16

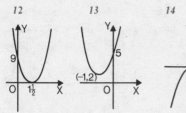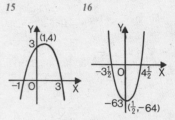

17a T   *b* F   *c* F   *d* T   *e* T

18 $\{x:1 < x < 5\}$ *19* $\{x:-1 \leqslant x \leqslant 3\}$ *20* $\{x:x < 2 \text{ or } x > 4\}$

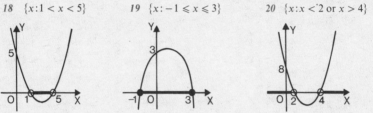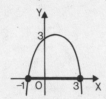

21 $\{x:x \leqslant \frac{1}{2} \text{ or } x \geqslant 2\frac{1}{2}\}$ *22* $\{x:-3 < x < 4\}$ *23* $\{x:x < -1.7 \text{ or } x > -0.3\}$

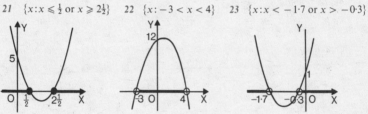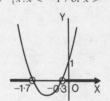

Page 15   *Exercise 6*

1 5, 6   *2* −5, 6   *3* −3, −1   *4* 2, −4   *5* $\frac{2}{3}, \frac{1}{3}$ *6* $\frac{3}{4}, -\frac{1}{2}$

7 $-p, q$ *8* $m/l, n/l$ *9* $-2h/a, b/a$ *10* 0, 5   *11* 3, 0   *12* $-2\frac{1}{2}, -\frac{1}{2}$

13a 2, 4   *b* $\frac{1}{2}$ *c* 8   *d* −4   *14a* $-\frac{1}{3}, \frac{2}{3}$ *b* $-\frac{1}{2}$ *c* $-1\frac{8}{9}$ *d* $-1\frac{5}{6}$

15a 6   *b* 8   *c* 34   *d* 2   *16a* p, q *b* $p^2 - 3q$ *c* p/q *d* $(p^2 - 2q)/q^2$

Page 16   *Exercise 7*

1 $m = 8$ *2* $m = 16; \frac{1}{4}, \frac{3}{4}$ *3* $m = 7; \frac{1}{2}, 3\frac{1}{2}$

4 a $x^2 - 7x + 12 = 0$ *b* $x^2 - x - 2 = 0$ *c* $x^2 + 5x + 6 = 0$ *d* $2x^2 + x - 1 = 0$
   e $9x^2 - 15x + 4 = 0$

5 a $x^2 - 7x + 10 = 0$ *b* $x^2 - 5x - 14 = 0$ *c* $x^2 - 2x - 1 = 0$ *d* $x^2 - 10x + 22 = 0$

6 $\{-\frac{1}{3}, 1\frac{1}{2}\}$ *7* $4x^2 - 8x - 3 = 0$ *8* $3x^2 + 8x + 2 = 0$ *9* $4x^2 - 10x + 9 = 0$

10 $4x^2 - 21x + 25 = 0$ *11* All are true.   *12* 49

260 **ANSWERS**

Page 17   *Exercise 8B*

1 $(x-10)(x+10)$ *2* $(x-\sqrt{2})(x+\sqrt{2})$ *3* $(x+1-\sqrt{2})(x+1+\sqrt{2})$

4 $(x+\frac{1}{2}-\frac{1}{2}\sqrt{5})(x+\frac{1}{2}+\frac{1}{2}\sqrt{5})$ *5* $4(x-\frac{1}{2}-\frac{1}{2}\sqrt{2})(x-\frac{1}{2}+\frac{1}{2}\sqrt{2})$

6 $2(x+\frac{1}{2}-\frac{1}{2}\sqrt{23})(x+\frac{1}{2}+\frac{1}{2}\sqrt{23})$ *7* $(2x-3)(x+5)$ *8* $3(x+1-\frac{1}{3}\sqrt{6})(x+1+\frac{1}{3}\sqrt{6})$

9 $7(x-\frac{2}{7}-\frac{4}{7}\sqrt{2})(x-\frac{2}{7}+\frac{4}{7}\sqrt{2})$

Algebra – Answers to Chapter 2

Page 20   *Exercise 1*

1 a two   *b* three   *c* four   *d* one   *e* zero

2 a x^3+x^2+x+1; three   *b* x^4+4x+1; four   *c* $-x^3+x$; three

   d $25x^2-10x+1$; two   *e* $-6x^2+13x-6$; two   *f* $x^3+6x^2+11x+6$; three

3 a -6 *b* 3 *c* 6 *d* 6   *4 a* $2x-7$ *b* $2x-9$

5 -3 *6* 20   *7* -1 *8* 0   *9* 31   *10* 10   *11* 0

12 35/27   *13* 283   *14* 4

Page 22   *Exercise 2*

1 36   *2* 19   *3* -6 *4* 14   *5* 186   *6* 116

7 45   *8* 1   *9* 61   *10* 16   *11* 0·64   *12* 4

Page 24   *Exercise 3*

1 $(7\times5)+1$ *2* $(12\times8)+4$ *3* $(5\times27)+1$ *4* $(9\times96)+3$

5 $(13\times41)+10$ *6* $(31\times66)+0$ *7* $(x-3)6+26$ *8* $(x+4)2-9$

9 $(x+2)(x+3)-2$ *10* $(x+5)(8x-44)+231$ *11* $(x-5)(6x+2)-5$

12 $(x-2)(x^2+4x+11)+28$ *13* $(x+3)(x^2+3x-6)+3$

14 $(x+1)(2x^2-6x+1)+8$ *15* 9,9   *16* 17,17   *17* 6,6   *18* 0,0

Page 26   *Exercise 4*

1 $2x+5,9$ *2* $x-1,-11$ *3* $3x-8,15$ *4* $x^2+8x+19,39$

5 $x^2+7x+32,164$ *6* $x^2-3,1$ *7* $3x^2+13x+32,97$

8 $3x^2-16x+53,-155$ *9* $x^2-5x+14,-60$ *10* $x^3-x^2,7$

11 $P(x)=(x+2)(x+1)(x-1)$ *12* $x+3$

Page 28   *Exercise 5*

1 -10 *2* -1 *3* 9   *4* -3 *5* 0   *6* 1

7 -14 *8* 203   *9* -693 *10* 55

Page 29   *Exercise 6*

1   2   2 -7 3   1   4   2   5   19   6   4

7   $x^2 + 2x - 3, 5$ 8 $x^2 + 3x - 4, 8$ 9 $x^2 - 2x + 3, -10$

10   $x^2 + 4x + 1, -2$ 11 $3x^2 + 5x + 7\frac{1}{2}, 23\frac{1}{2}$ 12 $2x - 1, 4$

13   $a = 2, 2x^3 - 5x^2 + 4x - 2$ 14 $a = 9, 3x^2 - 5x + 3$

Page 30   *Exercise 7*

6   *c*   7   *c*   8 $(x-1)(x-2)(x+3)$ 9 $(x-1)(x-3)(x-4)$

10   $(x-2)(x-5)(x+7)$ 11 $(x+1)(x+3)(2x-1)$ 12 $(t-1)(t+1)(3t-4)$

13   $(t-3)(2t^2 + t + 7)$ 14 -16 15   6   16 -4

17   $2; x-1$ and $x+1$ 18 $-8, 12; (x-1)(x+3)(x-2)(x+2)$

20a $(x-h)(x^2 + hx + h^2)$ *b   Hint.* $(x^3 - h^3) + (px^2 - ph^2) + (qx - qh)$

Page 32   *Exercise 8*

1   6, 2   2 $\frac{1}{2}, -\frac{2}{3}$ 3 $2\frac{1}{2}, 4$ 4 $84; -7, 4$ 5 $-3, -1, 2$

6   $-5, 1, 3$ 7 $-2, 1, 1$ 8 $-3, -2, 1$ 9 $-1, 1$ 10 $-3, -2, 1, 4$

11   $-\frac{3}{5}, 1, 1\frac{1}{2}$ 12 $-2; x^2 - x + 1 = 0$ has no real roots.

13   $-9; -\frac{1}{2}, 2, 3$ 14a $\{1, 2, 3\}$ *b* $(1, 0), (2, 0), (3, 0); 2, -1, 2$

15a $(-1, 0), (1, 0)$ *b* $(-1, 0),$ max.; $(\frac{1}{3}, -1\frac{5}{27}),$ min.   *c* $-1 < x < 1$ and $x < -1$

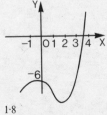
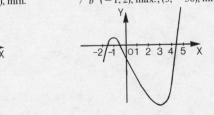

16a $(-1, 0), (2, 0)$ *b* $(-1, 0),$ max.; $(1, -4),$ min.   *c* $-1 < x < 2$ and $x < -1$

17   $\{30, 90, 150\}$ 18 $\{33{\cdot}7, 56{\cdot}3, 135, 213{\cdot}7, 236{\cdot}3, 315\}$

Page 34   *Exercise 9*

3   *b* $0{\cdot}4$ 4   *b* $1{\cdot}2$ 5 $2{\cdot}1$

6   *b* $(0, -6),$ max.; $(2, -10),$ min.   7   *b* $(-1, 2),$ max.; $(3, -30),$ min.

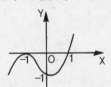
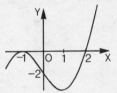

$1{\cdot}8$

Algebra – Answers to Chapter 3

Page 37   *Exercise 1*

1 a^5	*2* x^{16}	*3* p^4	*4* y^{50}	*5* k^5	*6* a^5
7 b	*8* x^5	*9* a^6	*10* a^6	*11* x^{100}	*12* z^{12}
13 x^3y^6	*14* $2^5 = 32$	*15* $5^3 = 125$	*16* $2^6 = 64$	*17* 3	*18* a^6b^6
19 x	*20* a^{12}	*21* 3	*22* $6x^5$	*23* $2a^3$	*24* $16a^8b^{12}$
25 4	*26* -8	*27* 1	*28* -1		

29a $(-1)^n = 1$, n even and $(-1)^n = -1$, n odd.

 b $f(x) = x^n - 1 \Rightarrow f(1) = 1^n - 1 = 1 - 1 = 0$ for all n. Result follows. Also, $f(-1) = (-1)^n - 1$ which is zero if n is even, and -2 if n is odd, so $x + 1$ is a factor of $x^n - 1$ only if n is even.

 c $x + 1$ a factor only if n is odd; $x - 1$ is not a factor.

30a x^2 *b* $-x^5$ *c* $-8x^3$ *d* $64x^6$

31 $(ab)^4 = ab \times ab \times ab \times ab = a \times a \times a \times a \times b \times b \times b \times b = a^4b^4$

32 $(ab)^p = ab \times ab \times \ldots$ to p factors $= (a \times a \times \ldots$ to p factors$) \times (b \times b \times \ldots$ to p factors$) = a^p \times b^p$

33 $(a^p)^q = a^p \times a^p \times \ldots$ to q factors $= (a \times a \times \ldots$ to p factors$) \times (a \times a \times \ldots$ to p factors$) \times \ldots$ to q factors $= a \times a \times \ldots$ to pq factors $= a^{pq}$

Page 39   *Exercise 2*

1 $\dfrac{1}{a^2}$	*2* $\dfrac{1}{a^{1/2}}$	*3* $\dfrac{1}{a^{2/3}}$	*4* a^3	*5* $\dfrac{2}{x}$	*6* $\dfrac{3}{x^{1/3}}$
7 $3a^2$	*8* $\frac{1}{2}a^{1/2}$	*9* $\frac{1}{9}$	*10* 4	*11* $\frac{1}{4}$	*12* 1
13 1	*14* 4	*15* 8	*16* 9	*17* 2	*18* $\frac{1}{2}$
19 4	*20* $\frac{1}{9}$	*21* $\frac{1}{4}$	*22* 8	*23* 24	*24* 3
25 a^2	*26* a	*27* a^2	*28* $a^{1/2}$	*29* 1	*30* 1
31 1	*32* 4	*33* $12a^{2/3}$	*34* $2c$	*35* $2x$	*36* $4z$
37 8	*38* 24	*39* 36	*40* $\frac{2}{9}$	*41* $\frac{3}{2}$	*42* 1
43 $\dfrac{1}{p}$	*44* $\dfrac{1}{x^2}$	*45* $\dfrac{1}{a^4}$	*46* $\dfrac{4}{t}$	*47* $\dfrac{1}{a}$	*48* $\dfrac{6}{x}$
49 $a+1$	*50* $x - \dfrac{1}{y}$	*51* $e^{2x} - 2 + \dfrac{1}{e^{2x}}$		*52* 400	*53* 5

Page 41   *Exercise 3*

1 See Figure 1.   *a* 0·7   *b* 5·7   *c* 2·7   *d* 3·3

2 Similar in shape to graph of $f : x \to 2^x$; (0, 1).   Yes.

3 a (*1*) 3·2 (*2*) 16 *b* (*1*) 1·6 (*2*) 1·4

4 straight line through (0, 1) parallel to *x*-axis

5 See Figure 2. *a* *y*-axis *b* (0, 1)

Page 44 *Exercise 4*

1 $\log_{10} \frac{1}{100} = -2, \log_{10} 1 = 0, \log_{10} 100 = 2, \log_{10} 10\,000 = 4$

2 a $\log_2 8 = 3$ *b* $\log_2 4 = 2$ *c* $\log_2 1 = 0$ *d* $\log_2 \frac{1}{16} = -4$ *e* $\log_2 p = q$

3 a $16 = 2^4$ *b* $128 = 2^7$ *c* $125 = 5^3$ *d* $\frac{1}{64} = 2^{-6}$ *e* $7 = 7^1$ *f* $a = 2^5$

  g $4 = 3^x$ *h* $\sqrt[3]{2} = 2^{1/3}$ *i* $n = a^b$

4 a $\log_3 9 = 2$ *b* $\log_4 64 = 3$ *c* $\log_3 \frac{1}{9} = -2$ *d* $\log_9 1 = 0$ *e* $\log_5 \sqrt[3]{5} = \frac{1}{3}$

  f $\log_{27} \frac{1}{3} = -\frac{1}{3}$ *5 a* 1 *b* $\frac{1}{16}$ *c* 32 *d* −3

6 a 2 *b* 1 *c* 0 *d* −2 *e* $\frac{1}{2}$

7 a 2·1 *b* 3·6 *c* 1·9 *d* 2·9 *e* 0·6

Page 45 *Exercise 5*

1 a 1 *b* 2 *c* 2   *2 a* 2 *b* 2 *c* 0 *d* −3

3 a F *b* T *c* T *d* T *e* F *f* F *g* T *h* F

5 a 1 *b* 1 *c* 3 *d* 0 *e* 3 *f* 2

6 a 2 *b* 3 *c* 2 *d* 10 *e* 4 *f* 9   *7* 4

9 $y = \log_e \dfrac{x}{1+x} \iff \dfrac{x}{1+x} = e^y$, etc.

Page 47 *Exercise 6*

1 a 2 *b* 1 *c* 0   *2 a* 0 *b* 0   *3 b* (*1*) $\log 3$ (*2*) $\frac{4}{3}$

4 a T *b* T *c* T *d* F *e* F *f* T *g* T *h* T   *5* none; each is equal to zero

6 a 0 *b* 0 *c* $\frac{1}{2}$ *d* 2   *7 a* 1·8 *b* 0·77 *c* 2·8 *d* 0·32

8 a 8 *b* 11   *9* 8   *10a* 5·65 *b* 30·1 *c* 1·06 *d* 0·585

11a 3^{11} *b* 15^6

13a $\log A = \log 4 + \log \pi + 2 \log r$ *b* $\log V = \log 1·33 + \log \pi + 3 \log r$

  c $\log T = \log 2\pi + \frac{1}{2}(\log l - \log g)$ *d* $\log p + 1·4 \log v = \log c$

Page 49 *Exercise 7*

1 2·8 *2* 2·3 *3* 1·5 *4* 0·68 *5* 1·1 *6* 1·4 *7* 1·5 *8* 0·65

13 1·84   *14a* −2·32 or $\bar{3}·68$ *b* −0·43 or $\bar{1}·57$

Page 52   *Exercise 8*

1 $y = 2x^3$ *2* $y = 2x^{0\cdot6}$ *3* $\dfrac{\log x \mid 0 \quad\;\; 0\cdot60 \;\; 0\cdot95 \;\; 1\cdot20}{\log y \mid 0\cdot60 \;\; 0\cdot30 \;\; 0\cdot11 \;\; 0}$; $y = 4x^{-0\cdot5}$

4 $E = 219p^{-1}$ *5* $E = 0\cdot5p^2$ *6* $m = 20d^2$

Page 55   *Exercise 9B*

1   1070   *2 a* 249   *b* 12 years   *3* 230   *4 a* 41   *b* 13·8 years

5   0·077   *6 b* 1·6   *7*   590 units   *8 a* 47 300   *b* 38 years

9 a 140°C   *b* 35°C   *10a.* £4050   *b* 4

Algebra – Answers to Chapter 4

Page 58   *Exercise 1*

1 a unary; numbers, matrices, vectors   *b* binary; numbers, matrices, vectors

 c binary; numbers, matrices   *d* unary; plane figures   *e* binary; sets

 f unary; sets   *g* binary; numbers, matrices, vectors

2 a number   *c* set   *e* another function or transformation

Page 59   *Exercise 2*

```
1   E O  2 D O  3 0 1 2 3..  4 0 1 2   5 I  R  R²   6 I  A  A²
    O E    O D    1 2 3 4..    1 2 0      R  R²  I      A  A²  I
                  2 3 4 5..    2 0 1      R² I  R      A²  I  A
                  3 4 5 6..
                  : : : :
```

```
7  1 2 3  4 ...   8  1 2 3 4   9  0 1 2 3   10  0   a      b      c    ...
   2 4 6  8 ...      2 4 1 3      1 2 3 0        a  2a   a⊕b  a⊕c  ...
   3 6 9 12 ...      3 1 4 2      2 3 0 1        b  b⊕a   2b   b⊕c  ...
   4 8 12 16 ...     4 3 2 1      3 0 1 2        c  c⊕a  c⊕b   2c   ...
   : : :  :                                     :  :     :     :
```

```
11  A B C D  12  P Q R S  13  φ A B C   14  φ   φ    φ   φ   15  f g h
    B A D C      Q R S P      A A C C       φ   A   {3}  A       g h f
    C D A B      R S P Q      B C B C       φ  {3}   B   B       h f g
    D C B A      S P Q R      C C C C       φ   A    B   C
```

Page 62   *Exercise 3*

1   All except *14*   *2* *1* E *2* D *3* 0 *4* 0 *5* I *6* I *7* I *8* I *9* 0 *10* 0 *11* A

 12 P *13* φ *14* C *15* f   *3*   All except *3, 7, 13, 14*

Page 63   *Exercise 4*

1 a e.g. $4-(3-2) = 3, (4-3)-2 = -1$ *b* e.g. $12 \div (3 \div 2) = 8, (12 \div 3) \div 2 = 2$

 c e.g. $3^{(1^2)} = 3, (3^1)^2 = 9$ *2 a* $p+(-q)+(-r)$, i.e. $(p-q)-r$ *b* none

 c $x^{(y^z)}$, since $(x^y)^z$ means x^{yz}

Page 63   *Exercise 5*

1   1, 2, 4, 5, 6, 8, 9, 10, 11, 12, 15   *2*   no; no inverses   *3*   yes

4   no; no inverses   *5*   no; 0 has no inverse   *6*   yes

7   no; not all functions have inverse functions   *8*   yes   *9*   yes   *10*   yes

11   no; no closure, no identity element   *12*   no; not all matrices have inverses

(when determinant = 0)   *13*   no; no inverses

Page 64   *Exercise 6*

1   yes; see Exercise 5, question *8*   *2*   no; see Exercise 5, question *13*

3   yes; see Exercise 5, question *9*   *4*   no; see Exercise 5, question *12*

Page 65   *Exercise 7*

1   *1* 2   *2* 2   *4* 3   *5* 3   *6* 3   *8* 4   *9* 4   *11* 4   *12* 4   *15* 3

2   e a b   *3* not possible   *4* all elements of the group are used up to fill
    a b e   the row; yes
    b e a

5   e a b c   e a b c   e a b c   e a b c
    a e c b   a e c b   a b c e   a c e b
    b c e a   b c a e   b c e a   b e c a
    c b a e   c b e a   c e a b   c b a e

6 a   yes   *b* Operation is not associative, e.g. $b*(d*c) \neq (b*d)*c$

Page 66   *Exercise 8*

1   I H X Y   *2* I H   *3* I H A B   *4* I M
    H I Y X   H I   H I B A   M I
    X Y I H   A B I H
    Y X H I   B A H I

5   I   R   R² A B C   *6* I   R   R² R³ X M₁ Y M₂   *7* all groups
    R   R² I C A B   R   R² R³ I M₂ X M₁ Y   except those in
    R² I   R B C A   R² R³ I R Y M₂ X M₁   questions *5*
    A   B   C I R R²   R³ I   R R² M₁ Y M₂ X   and *6*
    B   C   A R² I R   X M₁ Y M₂ I R R² R³
    C   A   B R R² I   M₁ Y M₂ X R³ I R R²
                                Y M₂ X M₁ R² R³ I R
                                M₂ X M₁ Y R R² R³ I

Page 69   *Exercise 9*

1 $0 \leftrightarrow I, 1 \leftrightarrow R, 2 \leftrightarrow R^2$ *2* $0 \leftrightarrow P, 1 \leftrightarrow Q, 2 \leftrightarrow R, 3 \leftrightarrow S$

3 a   1 3 5 7   *b* yes   *c* no; each element is its own inverse, which is not true for
    3 1 7 5   the group in question *9*
    5 7 1 3
    7 5 3 1   *4* $1 \leftrightarrow 0, 2 \leftrightarrow 1, 4 \leftrightarrow 2, 3 \leftrightarrow 3$

5 a *e a b c*   *b e a b c*   *e b a c*   *e a c b*
　　a e c b   *a b c e*   *b a c e*   *a c b e*
　　b c e a   *b c e a*   *a c e b*   *c b e a*
　　c b a e   *c e a b*   *c e b a*   *b e a c*

6　no; yes　　*7*　no one-to-one correspondence possible; yes, $-2 \leftrightarrow -1$, $0 \leftrightarrow 0$,

　　$2 \leftrightarrow 1$, etc.　　*8* isomorphic to group of symmetries of equilateral triangle

Algebra – Answers to Revision Exercises

Page 72　*Revision Exercise 1*

1 a $\{-\frac{1}{2}, \frac{1}{3}\}$　*b* $\{-0.45, 4.45\}$　*c* ϕ　**2** $h = 6$

3 a real, rational, distinct　*b* real, irrational　*c* not real

4 a (1) $k = 16$　(2) $k < 16$　　*b* (1) $k = 1$　　(2) $k \in R$, but $k \neq 1$

　c (1) $k = 1$ or 9 (2) $k < 1$ or $k > 9$　*d* (1) $k = \frac{1}{2}$ or $4\frac{1}{2}$ (2) $k < \frac{1}{2}$ or $k > 4\frac{1}{2}$

5 a $m = 0$ or $1\frac{1}{3}$　*b* $0 < m < 1\frac{1}{3}$　　*6* $y = 2x \pm 5$; $(-2, 1)$, $(2, -1)$

7　$y = \pm\sqrt{3x + 4}$　*8 a* Min T.P. $(2, -4)$ *b* Max T.P. $(-3, 16)$ *c* Min T.P. $(-1, 3)$

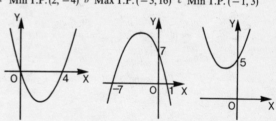

9 a (1) $x = \frac{1}{3}$ or 5
　　(2) $x < \frac{1}{3}$ or $x > 5$
　　(3) $\frac{1}{3} < x < 5$

　b (1) $x = -\frac{1}{4}$ or 1　(2) $-\frac{1}{4} < x < 1$　(3) $x < -\frac{1}{4}$ or $x > 1$

10a $r^2 - 2s$　*b* $-r/s$　*c* $(r^2 - 2s)/s$　*d* $r^2 - 4s$

11a $x^2 - 2x - 35 = 0$　*b* $12x^2 - 17x + 6 = 0$　*c* $x^2 - 4x + 1 = 0$

　d $mx^2 - (m^2 + 1)x + m = 0$　　*12a* $p = 0$　*b* $q = 1$

13 $\{-\frac{3}{4}, 1\frac{1}{3}\}$　　*14a* T　*b* F　*c* T　*d* F　*e* T　　*15a* $x^2 + (2 - k)x + (3 - k) = 0$

16a after 6 seconds　*b* after 1 second and 5 seconds　*c* no

Page 74　*Revision Exercise 2*

1 a $x^3 + x^2 - 3x - 3$; 3　*b* $4x^3 - 12x^2 + 9x$; 3　*c* $-x^4 - x^3 + x^2 + x$; 4

2 a 2, -13 *b* 3, 4　*3a* 1, -5, -1, -11　*b* $3x^2 - 8x - 3$; 0, 0

4　$2\frac{1}{4}$, 5·202　*5* $(x - 3)(x - 5) - 4$　*6a* $(x + 2)(x^2 + 5x - 13) + 35$

　b $(x - 3)(4x^2 + 12x - 3) + 2$　　*7 a* 13　*b* 187　*c* 4　　*8 a* 0　*b* -1　*c* 5

9    $x+1, x+3$ 10a $(x-1)(x+1)(x-4)$ b $(y-3)(y+3)(y^2+4)$

c $(z+3)(2z-1)(3z+2)$ d $(x-1)(x^2+2x+3)$ 11 -10

12   $-1, 2$ 13   6   15a $\{-\frac{1}{3}, 1, 3\}$ b $\{-5, 2, 3\}$ c $\{-2, -1, 1\frac{1}{2}\}$

d $\{-1, 0, 2\}$ 16 $1, -2$ 17a $(2x+3)(3x+2)(x-7)$

b $-1\frac{1}{2}, -\frac{2}{3}, 7$ 18 $(-1\frac{1}{2}, 0), (-\frac{1}{2}, 0), (2, 0)$ 20   0·4

21   $(-1, -19)$, min ; $(1, 13)$, max ; $(2, 8)$, min

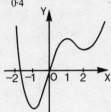

Page 76   Revision Exercise 3

1 a $3\frac{1}{4}$ b $\frac{7}{16}$ c 16   2 a $x-1$ b $x-1$ c $x-2x^2+x^3$

3 a F b F c T   5 b $(0, 1)$ c $x = 0$ d $7\frac{1}{2}$ 6 $\log_8 x$

7 a $\log_2 16 = 4$ b $\log_3 1 = 0$ c $\log_4 2 = \frac{1}{2}$ d $\log_{10} \frac{1}{100} = -2$

8 a $81 = 9^2$ b $16 = 4^2$ c $4 = 16^{1/2}$ d $\frac{1}{4} = 2^{-2}$

9 a 5 b 25 c 2 d 1 e 0 f -3 g 8 h 10

11b $(1, 0)$ c $y = 0$ d $\pm 1.5 (\pm 1.48)$ 12a 2 b -4 13 $\frac{1}{2}$

14a $\{-2, 4\}$ b $\{-1, \frac{1}{2}\}$ 15a $\frac{1}{2}$ b 4   16a $\log 2\sqrt{2}$ b $\log\sqrt{2}$ c 2

17   $y = 2x^3$ 18a 2·3 b 0·32 c 1·2   19   7

20a $2\left(1 - \dfrac{1}{2^n}\right)$; 2 b 11   21a 1·4 b 1·8 c 0·88 d 23

22   $m = n$ 23 $1.4, P \doteq 75/V^{1.4}$ 24   2·5, 3   25a 92 mg b 69 years

Page 78   Revision Exercise 4

1 a unary; numbers, vectors, matrices b binary; numbers

c binary; matrices d unary; sets

2 a closure, associativity, commutativity—yes; identity, 1; inverses—no (not a
Latin square) b no; a divides b

3 a no, no; 0 is identity, no (only 0 has an inverse); 0, yes; 0, yes; 0, yes

b 1 is identity, no; 1, no; 1, no; 1, no (0 has no inverse); 1, yes; 1, no; 1, yes

4 a no b no c yes d yes

5 a identity, E; inverse property does not hold; closure, associativity,
commutativity—yes b no c yes

6 *b*　All the group properties exist. yes

7 *a*　3 has no multiplicative inverse in Z　*b*　0 has no multiplicative inverse in R

　c　$\begin{pmatrix} 0 & 1 \\ 1 & 0 \end{pmatrix}$　*d*　no multiplicative inverse for $\begin{pmatrix} 1 & 1 \\ 1 & 1 \end{pmatrix}$　*e*　no inverse for $\{3, 4, 5\}$;

　solution set is $\left\{\{3, 4\} \cup Y : Y \text{ a subset of } \{1, 2, 6, 7, 8, 9\}\right\}$.

8 *a*　4　*b*　2　*c*　2　*d*　2　*e*　infinite group

9 *a*

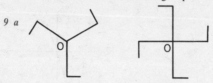

　b　Ls from O spaced at angles of $\dfrac{360°}{n}$ to each other

10*a*　$1 \leftrightarrow 0, 2 \leftrightarrow 1$　*b*　no; different orders　*c*　$0 \leftrightarrow 0, 2 \leftrightarrow 1, 2n \leftrightarrow n$, etc. (both infinite

　groups)　*d*　$1 \leftrightarrow I, 3 \leftrightarrow R, 9 \leftrightarrow R^2, 7 \leftrightarrow R^3$ *or* $1 \leftrightarrow I, 7 \leftrightarrow R, 9 \leftrightarrow R^2, 3 \leftrightarrow R^3$

　e　$\log_b x \leftrightarrow \log_a x$ *or* $\log_a \dfrac{1}{x}$

　f　no; one is non-commutative, the other is commutative

Algebra – Answers to Cumulative Revision Section

Page 83　Exercise 1

1　$(2, 4),\quad (3, -2),\quad (-4, -1)$　2　$\{(1, -1, 1)\}$　3　$\{(2, 0, -1)\}$　4　$\{(\tfrac{1}{2}, 1, 1\tfrac{1}{2})\}$

5　$2, 1, -3$　6　$x^2 + y^2 - 4x + 2y + 1 = 0$　7　$\{(0, -\tfrac{1}{2}), (-1, -1)\}$

8　$\{(-1, 2), (2\tfrac{3}{13}, -\tfrac{2}{13})\}$　9　$\{(2, 0), (-1\tfrac{1}{13}, -2\tfrac{4}{13})\}$　10　$(-1, 0), (1, 4)$

11　$(2, 0), (1\tfrac{1}{3}, -1\tfrac{3}{5})$　12　$(-2, 2), (6, 8)$

Page 84　Exercise 2

1　$125; 6$　3　$0, 5, 18; 0, 5, 13$　4　80　5　$1\tfrac{5}{27}$

6 *a*　$\tfrac{1}{2}\left(1 - \dfrac{1}{3^n}\right); \tfrac{1}{2}$　*b*　6　7 *a*　(1) is geometric, (3) is arithmetic.

　b　fifth terms: $9\sqrt{3}, 5\sqrt{3}$. nth terms: $(\sqrt{3})^n, n\sqrt{3}$　*c*　$\dfrac{\sqrt{3}[(\sqrt{3})^n - 1]}{\sqrt{3} - 1}; \tfrac{1}{2}n\sqrt{3}(1 + n)$

8 *a*　$\dfrac{1}{1 - q}, -1 < q < 1$　*b*　$\dfrac{1}{1 - q^2}$　9 *a*　Common difference is 2 in each case.

　b　$n^2, 2n + 1; n(n + 1), 2(n + 1)$　*c*　$10\,000, 201; 10\,100, 202$

11　£108, £116·64, £125·97; 1·08; £214

Page 86   Exercise 3

1 *a* $\begin{pmatrix} 0 & 0 \\ 2 & 0 \end{pmatrix}$ *b* $\begin{pmatrix} 4 & 6 \\ -4 & 0 \end{pmatrix}$ *c* $\begin{pmatrix} 12 & -3 \\ -7 & -7 \end{pmatrix}$ 2 $-2, 1, -3, \frac{1}{2}$

3 *a* $\begin{pmatrix} 4 & 0 \\ 4 & 0 \end{pmatrix}$ *b* $\begin{pmatrix} 1 & 0 \\ 1 & 0 \end{pmatrix}$ 4 $AB = \begin{pmatrix} -3 & 12 \\ -2 & 8 \end{pmatrix}$, $BC = (-13 \quad 2)$,

$BA = (5)$, $CA = \begin{pmatrix} -1 \\ -9 \end{pmatrix}$ 5 $-2, 0$ 7 $\begin{pmatrix} 1 & -1 \\ 2 & 3 \end{pmatrix}$; $(1, 7), (0, 20), (-1, -12)$

8   anticlockwise rotation of $30°$ 9 $\begin{pmatrix} 3 & -2 \\ -7 & 5 \end{pmatrix}$

10   $\sqrt{2}\begin{pmatrix} \frac{1}{2} & \frac{1}{\sqrt{2}} \\ -\frac{1}{2} & \frac{1}{\sqrt{2}} \end{pmatrix}$; $x = \frac{1}{\sqrt{2}}x' + y', y = -\frac{1}{\sqrt{2}}x' + y'$

11*a* $\begin{matrix} 1 & 1 & 0 \\ 0 & 1 & 1 \end{matrix}$ $\begin{matrix} 1 & 0 \\ 1 & 1 \\ 1 & 1 \end{matrix}$ *b* $\begin{pmatrix} 1 & 1 & 0 \\ 0 & 1 & 1 \end{pmatrix}$, $\begin{pmatrix} 1 & 0 \\ 1 & 1 \\ 1 & 1 \end{pmatrix}$, $\begin{pmatrix} 2 & 1 \\ 2 & 2 \end{pmatrix}$ *c* $\begin{pmatrix} 10 & 11 \\ 12 & 13 \end{pmatrix}$

Page 88   Exercise 4

1    $h^2 + h + 1, h^2 + 9h + 21, 8h + 20$ 2 $-1, 4; -5$

3 *a* $6, 5, -6$ *b* $-1\frac{1}{2}, \frac{2}{3}$ 4 *a* ± 5 *b* $-3, 1\frac{2}{3}$ *c* $0, -2$

5 *a* $2x^2 + 11x + 16; 2x^2 - x + 4; -1$ *b* $x + 6; 2x^2 - x + 7$

6 *a* $h \circ k$ *b* $h \circ h$ *c* $k \circ h$ *d* $k \circ k$ *e* $h \circ k \circ k$ *f* $k \circ h \circ h$

7   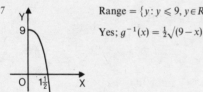   Range $= \{y : y \leqslant 9, y \in R\}$.

Yes; $g^{-1}(x) = \frac{1}{2}\sqrt{(9 - x)}$

8 *a* $\sqrt[3]{(x - 1)}; R$ *b* $\frac{1}{2}(2x + 3); R$ *c* $3(1 - x); R$ *d* $\dfrac{5x + 1}{2x}; R$, except $x = \frac{5}{2}$ and 0

e $\dfrac{3x - 2}{4x}; R$, except $x = \frac{3}{4}$ and 0 *f* $\dfrac{1}{\sqrt[3]{2(1 - x)}}; R$, except $x = 0$ and 1

9 *a* $\frac{1}{2}x$ *b* $\dfrac{2}{3x}$ *c* $\dfrac{1}{3x}$ *d* $\dfrac{4}{3x}$ 10 $-2x - 2$

Algebra – Answers to Cumulative Revision Exercises

Page 90   Cumulative Revision Exercise 1

1   2·16   2 $\{x : 1 < x < 2\}$ 3   2500; 50   4 *a* F   *b* F   *c* F

5 *a* I, II, III   *b* I, VI, VII, IX   *c* I   *d* IV, V, VIII

6 a (1) 0 (2) 12 *b* 0·35   7 $\{x:x \geqslant 8\}$

8 a $(p+q-r)(p+q+r)$ *b* $(2x-5)(x+2)$ *c* $b(a+b)(a-3b)$

9 a $\{-1\frac{1}{2}, 2\}$ *b* $\{-1·4, 1·9\}$

Page 91   Cumulative Revision Exercise 2

1 a y *b* $x+2+\dfrac{1}{x}$ *c* 0   *2 a* $\{x:-5 < x < 5\}$ *b* $\{x:2 \leqslant x \leqslant 4\}$

  c $\{x:x < -2 \text{ or } x > 3\}$ *3 a* $(10-x)$ cm, $(10-2x)$ cm *c* 25, 5

4 $\{-2·6, -0·4\}$ *5 a* 6 *b* $4\sqrt{2}$ *c* $24\sqrt{2}$ *d* 1

6 a (1) I, II, VI, VII (2) I, II *b* $\{(x,y):y \leqslant 0, y \leqslant x+3, \text{ and } y \leqslant -x+3\}$

7 a $d = \frac{1}{15}v^2 + \frac{2}{3}v$ *b* 165, 20   *8 a* $\begin{pmatrix} 0 & -1 \\ 1 & 0 \end{pmatrix}, \begin{pmatrix} 0 & 1 \\ 1 & 0 \end{pmatrix}, \begin{pmatrix} -1 & 0 \\ 0 & 1 \end{pmatrix}, \begin{pmatrix} 1 & 0 \\ 0 & -1 \end{pmatrix}$

  b reflection in *y*-axis, and in *x*-axis *c* $\begin{pmatrix} -1 & 0 \\ 0 & -1 \end{pmatrix}$, half turn about origin

Page 92   Cumulative Revision Exercise 3

1 a $\{(-4, -3), (3, 4)\}$ *b* $\pm 5\sqrt{2}$ *2 a* 570 *b* only 384 *c* 4

3 *a* and *c*   *4 a* $-24, -17$ *b* $-4x^2+4x, 1-2x^2$ *c* $-1, 2$

5 a (1) $(x-1)(x-4)(x+2)$ (2) $(x+1)(x+3)(x-4)$ *b* (1) $\{-3, -1, \frac{1}{2}\}$
  (2) $\{-4, -3, -1, 1\}$

6 a T *b* F *c* F *d* T   *7 a* $n = \dfrac{Ir}{E-IR}$ *b* $n = \dfrac{km}{m-k}$ *c* $n = \pm\dfrac{1}{\sqrt{(r^2+1)}}$

Page 93   Cumulative Revision Exercise 4

1 $(x+1)(x+3)(x-5); (-1, 0), (-3, 0), (5, 0); -12, 16, 48$

2 a $\dfrac{1}{2}\left(1-\dfrac{1}{2^n}\right); \frac{1}{2}$ *b* 9   *3 a* (1) $\frac{3}{4}, -6, -8$ (2) ± 2

  b $1\frac{1}{2}, -5; 1$ *4 a* $1, -1; \begin{pmatrix} 1 & 0 \\ -3 & -3 \end{pmatrix}$ *b* $\begin{pmatrix} \frac{1}{3} & 1\frac{1}{3} \\ -2 & -5 \end{pmatrix}$

5 a 1 *b* 2·11 *c* (4)

6 a If a quadrilateral has exactly one axis of symmetry, it is a kite. T, F

  b If $-1 \leqslant \cos a° < 0$, then $810 < a < 990$. T, F

  c If $x \notin R$, then $x^2+2x+3 = 0$. T, F

  d If $0 < p < 1$, then $\log_{10} p < 0$. T, T

Page 94   ***Cumulative Revision Exercise 5***

1 a $\{(-1,3),(1,-2)\}$ *b* $\{(2,-1,3)\}$ *2 a* (*1*) $3-10x$ (*2*) $\frac{1}{10}(3-x)$

(*3*) $\frac{1}{2}x$ (*4*) $\frac{1}{5}(3-x)$ (*5*) $\frac{1}{10}(3-x)$ *b* $(g\circ f)^{-1}=f^{-1}\circ g^{-1}$

3 a yes *b* no *c* no   *4 a* $(x-1)(x-1)(x+3)$; $1,-3$

  b 0 (minimum), and $9\frac{13}{27}$ (maximum)   *5*   20

6 a -4 *b* 3 *c* 12 *d* $-3\frac{1}{2}$ *7* -4

8 a $\{x:-1<x<6\}$ *b* $\{x:-1\leqslant x\leqslant 2\}$ *c* $\{x:x<-4\text{ or }x>4\}$

Page 95   ***Cumulative Revision Exercise 6***

1 a $\{(2,5)\}$ *b* $\{(4,-3)\}$ *c* $\{[(pr-qs)/(p^2-q^2),(ps-qr)/(p^2-q^2)]\}$

2 a $3,-\frac{1}{2}$ *b* 0·7   *3 a* (*1*) $f^{-1}(x)=8-x$ (*2*) $f^{-1}(x)=\frac{1}{8}(x-1)$

(*3*) $f^{-1}(x)=\frac{1}{3}(2x+4)$ *b* (*1*) T (*2*) T (*3*) T (*4*) F (*5*) T (*6*) T   *4* 2, 4

5 $2,-2,-32$; $x^2+y^2+4x-4y-32=0$ *6 a* (*1*) $k=2$ (*2*) $k>2$ *b* (*2*)

7 $e^{-Rt/L}=1-\dfrac{IR}{E}$; $t=-L\log\left(1-\dfrac{IR}{E}\right)\Big/R\log e$; $t\doteqdot 0\cdot01$

Geometry – Answers to Chapter 1

Page 99   *Exercise 1*

1 a   *b*   *c*

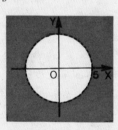

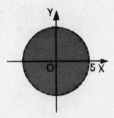

2 a $x^2+y^2=9$ *b* $x^2+y^2>9$ *c* $x^2+y^2<9$

3 a circle, centre O, radius a *b* region outside circle in *a*

  c region inside circle in *a*

4 a $x^2+y^2=r^2$ *b* $x^2+y^2>r^2$ *c* $x^2+y^2<r^2$

Page 101   *Exercise 2*

1 a $x^2+y^2=9$·   *b* $x^2+y^2=25$ *c* $x^2+y^2=64$ *d* $x^2+y^2=81$

  e $x^2+y^2=1\cdot44$ *f* $x^2+y^2=a^2$

2 a $x^2+y^2=13$ *b* $x^2+y^2=5$ *c* $x^2+y^2=16$ *d* $x^2+y^2=100$

3 a $(0, 0), 6$ b $(0, 0), 2\sqrt{3}$ c $(0, 0), \frac{3}{2}$

4   $x^2 + y^2 = 100$ 5 a $x^2 + y^2 = 16$ b $x^2 + y^2 = 32$

6 a 3 or -3 b 3 or -3 c 4 or -4 d 5 or -5

7 a Q b R c Q d P   8 a T b F c T d T e F

9   10π square units   10 $x^2 + y^2 = 9$; circle, centre O, radius 3

11a $x^2 + y^2 = 4$; circle, centre O, radius 2 b $x^2 + y^2 = 16$; circle, centre O, radius 4

c $x^2 + y^2 = 25$; circle, centre O, radius 5 d $x^2 + y^2 = 18$; circle, centre O, radius $3\sqrt{2}$

Page 102   Exercise 3

1 a on circle, centre $(5, 6)$, radius 3 b outside circle in a

c inside circle in a 2 a $(x-8)^2 + (y-3)^2 = 25$

b $(x-8)^2 + (y-3)^2 > 25$ c $(x-8)^2 + (y-3)^2 < 25$

3 a circle, centre (a, b), radius c b region outside circle in a

c region inside circle in a

4 a $(x-a)^2 + (y-b)^2 = r^2$ b $(x-a)^2 + (y-b)^2 > r^2$ c $(x-a)^2 + (y-b)^2 < r^2$

Page 103   Exercise 4

1 a $(x-1)^2 + (y-1)^2 = 9$ b $x^2 + (y-3)^2 = 16$ c $(x-5)^2 + y^2 = 4$

d $(x+5)^2 + (y-2)^2 = 49$ e $(x+3)^2 + (y+4)^2 = 36$ f $(x-2)^2 + (y+3)^2 = 1$

g $(x+1)^2 + y^2 = 36$ h $(x+6)^2 + (y+6)^2 = 9$

2 a $(x-6)^2 + (y-8)^2 = 100$ b $(x+8)^2 + (y-15)^2 = 289$ c $(x-1)^2 + (y-1)^2 = 8$

d $(x+2)^2 + y^2 = 41$ 3 a $(1, 3), 5$ b $(2, -4)$ 7

c $(-1, -8), 3$ d $(-5, 0), 1$ 4 $(x+2)^2 + (y-4)^2 = 100$

5   $(3, 3), (x-3)^2 + (y-3)^2 = 9$; $(3, -3), (x-3)^2 + (y+3)^2 = 9$;

$(-3, 3), (x+3)^2 + (y-3)^2 = 9$; $(-3, -3), (x+3)^2 + (y+3)^2 = 9$

6 a on b inside c outside d on   7 a 1:2 b 1:4

8 a $(x-1)^2 + (y-5)^2 = 25$ b (1) $(x-1)^2 + (y+5)^2 = 25$

(2) $(x+1)^2 + (y-5)^2 = 25$ (3) $(x+1)^2 + (y+5)^2 = 25$ (4) $(x-5)^2 + (y-1)^2 = 25$

9 a $(x-2)^2 + (y+1)^2 = 4$ b (1) $(x-2)^2 + (y-1)^2 = 4$

(2) $(x+2)^2 + (y+1)^2 = 4$ (3) $(x+2)^2 + (y-1)^2 = 4$ (4) $(x-1)^2 + (y+2)^2 = 4$

10a $(x-2)^2 + (y-3)^2 = 13$ b $(x+1)^2 + (y-2)^2 = 13$

11   $(x-3)^2 + (y-4)^2 = 8$ 12 $(y-y_1)/(x-x_1), (y-y_2)/(x-x_2)$

Page 105   Exercise 5

1   (1, 3), 2   *2* (2, 2), $2\sqrt{2}$;   *3* (3, 1), 1;   *4* (−1, −1), 1;
                          through origin   touches *x*-axis   touches both axes

         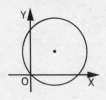

5 (−3, 0), 3;   *6* (0, 2), 2;   *7* (3, 4), 5;   *8* (−*a*, 0), *a*;
     touches *y*-axis   touches *x*-axis   through origin   touches *y*-axis
     at O   at O   at O

      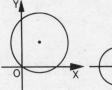

Page 106   Exercise 6

1 *a* (1, 3), 5   *b* (−2, −1), 2   *c* (2, −4), 5   *d* (2, 2), 1

2 *a* (2, −1), $\frac{3}{2}$ *b* (1, −$\frac{3}{4}$), $\frac{5}{4}$ *c* (cos θ, sin θ), cos θ

3 *a* (−4, 1), 5   *b* $2\sqrt{5}$ *4* $2\sqrt{2}$ *5* *b, c, e*

6   −3 or −10   *7* −1 or 6   *8* *a, c, d, e*   *9* −6   *10* −1

11   −5, 0, 0   *12* *a* $x^2 + y^2 - 6x - 8y = 0$ *b* $x^2 + y^2 - 6x - 8y + 5 = 0$

13 $x^2 + y^2 + 6x - 6y - 7 = 0$ *14* $x^2 + y^2 - 2x - 2y + 1 = 0$

15 $x^2 + y^2 - 12x + 10y - 11 = 0$; (6, −5), $6\sqrt{2}$

Page 107   Exercise 7

1

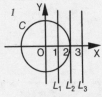

The line $x = 2$

2 *a* $\{(2\sqrt{2}, 1), (−2\sqrt{2}, 1)\}$ *b* $\{(\sqrt{5}, 2), (−\sqrt{5}, 2)\}$

c $\{(0, 3)\}$ $\{ \}$.

The lines $y = 1$ and $y = 2$ cut the circle in two points,
$y = 3$ in one point, $y = 4$ in none.

3 *a* (±2, 0), (0, ±2)   *b* (±6, 0), (0, ±6)   *c* (±8, 0), (0, ±8)

4 *a* (±2, ±2), (±2, ∓2)   *b* (±4, ±4), (±4, ∓4)   *c* (±7, ±7), (±7, ∓7)

5   The line does not cut the circle.

6 *a* $(\pm 2, \pm 4)$ *b* $(0, -4), (4, 0)$ *c* $(2, 2)$

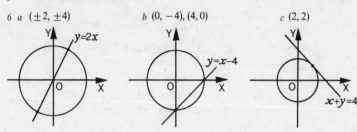

7 *a* $(-2, -1), (-1, 0)$ *b* $(-4, 0)$ 9 $x^2 + y^2 - 4x - 2y = 0; (0, 2), (1, 3)$

10    $5; x^2 + y^2 - 10x - 6y + 9 = 0; 8$

Page 109   Exercise 8

1    $4y = 3x + 25$ 2 $3y = x - 10$ 3 $3y = 2x + 13$

4 *a* $y = -x + 4$ *b* $y = x - 4$ *c* $y = -x - 4$ *d* $y = x + 4$

5    $y = 2$ 6 $3y = 7x - 31$ 7 $y = 4x + 6$

8    $4y = -3x + 25, 4y = 3x - 25; (8\frac{1}{3}, 0)$ 9 $(0, 0), (0, 6), (8, 0); 4x + 3y = 0,$
     $3y = 4x - 32, 3y = 4x + 18; (8, 6)$

Page 111   Exercise 9

1    $(2, 0)$ 2 $(-2, 4)$ 4 5   5 *a* $\frac{1}{2}; y = -2x + 10$

6 *a* $y = -3x - 10, (-6, 8)$ *b* $y = 3x - 10, (6, 8)$ 7 *b* (1) $0, \frac{4}{3}$ (2) $0 < k < \frac{4}{3}$
     (3) $k < 0$ or $k > \frac{4}{3}$ 8 *b* (1) $\frac{1}{3}, -3$ (2) $-3 < k < \frac{1}{3}$ (3) $k < -3$ or $k > \frac{1}{3}$

9    $y = mx + 2$ 10 $3y = 4x + 15, 3y = -4x + 15$

11    $x^2 + y^2 = 6$; a circle, centre O, radius 4; $4y = 3x - 20, 4y = -3x - 20$

12    $14, -6$ 13*a* $x^2 + y^2 - 8x - 6y = 0$ *b* $(0, 6), (4, -2)$ 14 $\sqrt{10}$

Geometry – Answers to Chapter 2

Page 115   Exercise 1

1 *a* reflection in $y = x$ *b* reflection in $y = -x$ 2 *a* $\begin{pmatrix} -1 & 0 \\ 0 & 1 \end{pmatrix}$ *b* $\begin{pmatrix} 0 & 1 \\ -1 & 0 \end{pmatrix}$

3 | Transformation | Matrix | Transformation | Matrix |
|---|---|---|---|
| Identity | $\begin{pmatrix} 1 & 0 \\ 0 & 1 \end{pmatrix}$ | Reflection in $y = x$ | $\begin{pmatrix} 0 & 1 \\ 1 & 0 \end{pmatrix}$ |
| Half turn about O | $\begin{pmatrix} -1 & 0 \\ 0 & -1 \end{pmatrix}$ | Reflection in $y = -x$ | $\begin{pmatrix} 0 & -1 \\ -1 & 0 \end{pmatrix}$ |

Reflection in x-axis $\begin{pmatrix} 1 & 0 \\ 0 & -1 \end{pmatrix}$ Rotation about O of $\frac{1}{2}\pi$ radians $\begin{pmatrix} 0 & -1 \\ 1 & 0 \end{pmatrix}$

Reflection in y-axis $\begin{pmatrix} -1 & 0 \\ 0 & 1 \end{pmatrix}$ Rotation about O of $-\frac{1}{2}\pi$ radians $\begin{pmatrix} 0 & 1 \\ -1 & 0 \end{pmatrix}$

4 a $(3, 1), (-2, 3), (4, 6)$ b $(-3, 1), (2, 3), (-4, 6)$

5 a $(1, -1), (-2, 0), (-3, 3), (0, 2)$ b $(1, 1), (0, -2), (-3, -3), (-2, 0)$

6 a $\begin{pmatrix} 1 & 0 \\ 0 & -1 \end{pmatrix}$, reflection in x-axis b $\{(1, -2), (5, 10), (a, -b), (c, -d)\}$

7 a $(0, 0), (2, 0), (4, 1), (2, 1)$ c no d yes

8 a $\begin{pmatrix} 3 & -4 \\ 4 & 3 \end{pmatrix}$ b $(0, 0), (3, 4), (-1, 7), (-4, 3)$ c dilatation and rotation

9 a $(4, -2), (-2, 16)$ b $(1, 7)$ in each case   10 $1, 3$ 11 $2, 1, -1, 3$

Page 117   Exercise 2

1 a $(3a, 3b)$ b $\begin{pmatrix} 3a \\ 3b \end{pmatrix}$ 2 a $(-2a, -2b)$ b $\begin{pmatrix} -2 & 0 \\ 0 & -2 \end{pmatrix}$ 3 $\begin{pmatrix} k & 0 \\ 0 & k \end{pmatrix}$

4 a $\{(2, 6), (-4, 8), (6, -10), (2a, 2b)\}$ b $\{(k, 3k), (-2k, 4k), (3k, -5k), (ka, kb)\}$

5 $\begin{pmatrix} 2 & 0 \\ 0 & 2 \end{pmatrix}$; $(2, 6), (8, 6), (8, 12), (2, 12)$

6 a $\begin{pmatrix} 2 & 0 \\ 0 & 2 \end{pmatrix}$ b $\begin{pmatrix} -\frac{1}{2} & 0 \\ 0 & -\frac{1}{2} \end{pmatrix}$ c $\begin{pmatrix} -p & 0 \\ 0 & -p \end{pmatrix}$

7 $(4, 2), (10, 2), (10, 6), (4, 6)$; dilatation $[O, 2]$ and reflection in x-axis

8 $(-3, 6), (-3, 15), (-9, 15), (-9, 6)$; dilatation $[O, 3]$ and reflection in $y = x$

Page 119   Exercise 3

1 a π b $\frac{1}{2}\pi$ 2 a $\begin{pmatrix} \frac{1}{2}\sqrt{3} & -\frac{1}{2} \\ \frac{1}{2} & \frac{1}{2}\sqrt{3} \end{pmatrix}$ b $\begin{pmatrix} \frac{1}{2} & -\frac{1}{2}\sqrt{3} \\ \frac{1}{2}\sqrt{3} & \frac{1}{2} \end{pmatrix}$ c $\begin{pmatrix} -\frac{1}{2} & -\frac{1}{2}\sqrt{3} \\ \frac{1}{2}\sqrt{3} & -\frac{1}{2} \end{pmatrix}$

3 $(\sqrt{2}, \sqrt{2}), (-\sqrt{2}, \sqrt{2}), (-\sqrt{2}, -\sqrt{2}), (\sqrt{2}, -\sqrt{2})$ 4 $(-2, 2\sqrt{3}), (-2, -2\sqrt{3})$

5 $(-\sqrt{3}, -1), (\sqrt{3}, -1), (0, 2)$. Equilateral triangle maps onto itself under rotation about O of $\frac{2}{3}\pi$ radians.

6 a If $\cos\theta = \frac{3}{5}$, $\sin\theta = \pm\frac{4}{5}$ b $(1, 7), (-3, 4)$ c $(0, 0), (0, 5), (-5, 5), (-5, 0)$

7 a $\begin{pmatrix} \frac{1}{2} & -\frac{1}{2}\sqrt{3} \\ \frac{1}{2}\sqrt{3} & \frac{1}{2} \end{pmatrix}$ b $(1, \sqrt{3}), (-1, \sqrt{3}), (-2, 0), (-1, -\sqrt{3}), (1, -\sqrt{3}), (2, 0)$.
   c Regular hexagon has sixth-order rotational symmetry.

Page 121   Exercise 4

1 a half turn about O *b* $(-a, -b)$ *c* $\begin{pmatrix} 1 & 0 \\ 0 & -1 \end{pmatrix}, \begin{pmatrix} -1 & 0 \\ 0 & 1 \end{pmatrix}$ *d* $\begin{pmatrix} -1 & 0 \\ 0 & -1 \end{pmatrix}$,

$\begin{pmatrix} -a \\ -b \end{pmatrix}$ *2 a* a half turn about O *b* $(-a, -b)$ *c* $\begin{pmatrix} 0 & 1 \\ 1 & 0 \end{pmatrix}, \begin{pmatrix} 0 & -1 \\ -1 & 0 \end{pmatrix}$

d Sense of rotation of half turn does not matter.   *3 a* rotation about O of

$\frac{1}{2}\pi$ radians; $(-b, a)$ *b* $\begin{pmatrix} 1 & 0 \\ 0 & -1 \end{pmatrix}, \begin{pmatrix} 0 & 1 \\ 1 & 0 \end{pmatrix}$ *c* rotation about O of $-\frac{1}{2}\pi$ radians

4 $\begin{pmatrix} 1 & 0 \\ 0 & -1 \end{pmatrix}$, reflection in *x*-axis   *5 b* $k\begin{pmatrix} 1 & 0 \\ 0 & 1 \end{pmatrix}\begin{pmatrix} a & b \\ c & d \end{pmatrix} = k\begin{pmatrix} a & b \\ c & d \end{pmatrix}\begin{pmatrix} 1 & 0 \\ 0 & 1 \end{pmatrix}$

6 a reflection in *x*-axis; reflection in *y*-axis   *c*

∘	I	P	Q	H
I	I	P	Q	H
P	P	I	H	Q
Q	Q	H	I	P
H	H	Q	P	I

d yes

7 $\begin{pmatrix} \frac{3}{5} & -\frac{4}{5} \\ \frac{4}{5} & \frac{3}{5} \end{pmatrix}\begin{pmatrix} 5 & 0 \\ 0 & 5 \end{pmatrix} = \begin{pmatrix} 3 & -4 \\ 4 & 3 \end{pmatrix}$; yes

8 a Hint: $\cos\theta_1 \cos\theta_2 - \sin\theta_1 \sin\theta_2 = \cos(\theta_1 + \theta_2)$,
$\sin\theta_1 \cos\theta_2 + \cos\theta_1 \sin\theta_2 = \sin(\theta_1 + \theta_2)$

b Hint: $\cos^2\theta - \sin^2\theta = \cos 2\theta$, $2\sin\theta\cos\theta = \sin 2\theta$

9 a $r\cos\theta, r\sin\theta$ *d* $\begin{pmatrix} \cos 2\alpha & \sin 2\alpha \\ \sin 2\alpha & -\cos 2\alpha \end{pmatrix}$ *e* $\begin{pmatrix} \cos 2\beta & \sin 2\beta \\ \sin 2\beta & -\cos 2\beta \end{pmatrix}$

10a $\begin{pmatrix} 1 & 2 \\ 1 & -3 \end{pmatrix}, \begin{pmatrix} 1 & 0 \\ 0 & -1 \end{pmatrix}, \begin{pmatrix} 1 & 2 \\ -1 & 3 \end{pmatrix}$ *b* $\{(0,0), (4,-4), (12,3), (8,7)\}$

11a rotation about O through θ radians where $\cos\theta = \frac{4}{5}$ and $\sin\theta = \frac{3}{5}$, and dilatation
$[O, 5]$ *b* $\begin{pmatrix} 4 & -3 \\ 3 & 4 \end{pmatrix}$

12b $\begin{pmatrix} 0 & -1 \\ 1 & 0 \end{pmatrix}$; rotation about O of $\frac{1}{2}\pi$ radians *c* half turn about O

Page 125   Exercise 5

1 a $\overrightarrow{AD}$ *b* $\overrightarrow{BG}$ *c* $\overrightarrow{AC}$ *d* $\overrightarrow{BG}$ *e* $\mathbf{O}$ *f* $\overrightarrow{DC}$

2 a CR *b* BS *c* translation *d* AP∥BS

3 a $p = -4, q = 2$ *b* $p = 5, q = -1$ *c* $p = 11, q = 0$ *d* $p = 2, q = 3$

4 a $(10, -4)$ *b* $(-3, 14)$ *c* $h = 2$ *d* $[a + 2(k-h), b]$

5

∘	I	M_1	M_2	H
I	I	M_1	M_2	H
M_1	M_1	I	H	M_2
M_2	M_2	H	I	M_1
H	H	M_2	M_1	I

M_1, M_2, H

6 *a* rotation of $\frac{1}{2}\pi$ radians about A in sense AC to AB

b three times *c* square

7 *a* (2, 0) *b* $(\sqrt{2}, \sqrt{2})$ *c* $(-2\sqrt{2}, 2\sqrt{2}); (0, -2), (\sqrt{2}, -\sqrt{2}), (2\sqrt{2}, 2\sqrt{2})$

8 *a*

∘	D	O
D	D	O
O	O	D

Identity element is *D*.

b

∘	E	O
E	E	O
O	O	E

similar pattern of entries

Page 127 Exercise 6

1

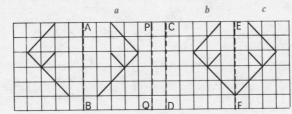

translation $\begin{pmatrix} 12 \\ 0 \end{pmatrix}$; reflection in PQ

2 *a* (1) (1, 2) (2) (5, 2) (3) (9, 2); translation $\begin{pmatrix} 2 \\ 0 \end{pmatrix}$ *b* (8, 4) *c* (12 − *a*, *b*)

3 *a* translation *b* reflection   4

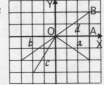

$M_4 \circ M_3 \circ M_2 \circ M_1 = I$

5 *a* $\begin{pmatrix} 1 & 0 \\ 0 & -1 \end{pmatrix}$ *b* $\begin{pmatrix} -1 & 0 \\ 0 & -1 \end{pmatrix}$

c $\begin{pmatrix} 0 & -1 \\ -1 & 0 \end{pmatrix}$ *d* DCBA = I   6 *n* even, rotation; *n* odd, reflection

7 *D* or *O* according to whether the number of *O*s is even or odd.

8 a translation; sum of separate translations

9 a rotation; sum of separate rotations

10 a reflection in the line making angle $\frac{1}{2}\theta$ radians with the given axis of reflection

Page 131 Exercise 7

1 Reflections in: *a* x-axis *b* y-axis *c* y = x *d* y = −x

Two successive reflections in the same axis restore the original configuration.

2 *a* $\begin{pmatrix} \cos\theta & -\sin\theta \\ \sin\theta & \cos\theta \end{pmatrix}$ *b* $\begin{pmatrix} \cos\theta & \sin\theta \\ -\sin\theta & \cos\theta \end{pmatrix}$ *c* rotation, −θ radians about O

3 *a* $(1,3),(3,7),(0,6)$ *b* $\frac{1}{2}\begin{pmatrix} 1 & 1 \\ -1 & 1 \end{pmatrix}$ 4 $\frac{1}{3}\begin{pmatrix} 1 & -1 \\ 1 & 2 \end{pmatrix}$; $(1,1),(6,2),(7,4),(2,3)$

5 *a* $\begin{pmatrix} 1 & 2 \\ 2 & -1 \end{pmatrix}$ *b* $\frac{1}{5}\begin{pmatrix} 1 & 2 \\ 2 & -1 \end{pmatrix}$ 6 *a* Plane maps to *x*-axis. *b* no

   c (*1*) not a one-to-one correspondence of points (*2*) no inverse matrix

7 *a* $\begin{pmatrix} 3 & 0 \\ 0 & 3 \end{pmatrix}$ *b* $\frac{1}{9}\begin{pmatrix} 3 & 0 \\ 0 & 3 \end{pmatrix}$ 8 *a* reflection in *x*-axis, reflection in $y = x$, rotation
      of $\frac{1}{2}\pi$ about O, rotation of $-\frac{1}{2}\pi$ about O

   b $A^{-1}B^{-1}$ *c* Second transformation must be 'undone' first. 9 *a* I

   b $B^{-1}A^{-1}$ 10*a* $\begin{pmatrix} 7 & -7 \\ 1 & -5 \end{pmatrix}$ *b* $\frac{1}{28}\begin{pmatrix} 5 & -7 \\ 1 & -7 \end{pmatrix}$

11*a* P, reflection in *y*-axis; Q, rotation of $-\frac{1}{2}\pi$ radians about O

   b $\begin{pmatrix} 0 & -1 \\ -1 & 0 \end{pmatrix}$, reflection in $y = -x$ *c* $\begin{pmatrix} 0 & 1 \\ 1 & 0 \end{pmatrix}$, reflection in $y = x$

   d $T^{-1} = \begin{pmatrix} 0 & -1 \\ -1 & 0 \end{pmatrix}$, $S^{-1} = \begin{pmatrix} 0 & 1 \\ 1 & 0 \end{pmatrix}$; successive reflections in the same axis

      restore the original configuration.

Page 133 Exercise 8

1 *a* $\begin{pmatrix} a_1 \\ b_1 \end{pmatrix} = \begin{pmatrix} 1 & 1 \\ 0 & 1 \end{pmatrix}\begin{pmatrix} a \\ b \end{pmatrix}$ *b* $\begin{pmatrix} a \\ b \end{pmatrix} = \begin{pmatrix} 1 & -1 \\ 0 & 1 \end{pmatrix}\begin{pmatrix} a_1 \\ b_1 \end{pmatrix} = \begin{pmatrix} a_1 - b_1 \\ b_1 \end{pmatrix}$; $a = a_1 - b_1$,

   $b = b_1$ *c* $x + 1 = 0$

2 *a* $y = -2x$ *b* $x - 3y + 1 = 0$ *c* $x + 1 = 0$ *d* $x - y = 3$

3 *a* $2x + 3y - 4 = 0$ *b* $3x + 2y - 4 = 0$ *c* $16x - 19y + 8 = 0$

4 *a* $3x + y - 2 = 0$ *b* $3x - 2y + 2 = 0$

5 *a* $a = 3, b = 1$ *b* $\begin{pmatrix} 4 & 0 \\ 2 & 2 \end{pmatrix} = \begin{pmatrix} 2 & 0 \\ 0 & 2 \end{pmatrix}\begin{pmatrix} 2 & 0 \\ 1 & 1 \end{pmatrix}$ *c* $3x - 2y - 4 = 0$

6 *a* $3x + y = 3\sqrt{2}$ *b* $y + \sqrt{2} = 0$

7 *a* $x^2 - 2xy + 2y^2 = 9$

   b $(3, 0), (-3, 0), (3, 3), (-3, -3)$ *c*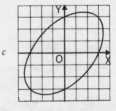

8 *a* $x^2 + y^2 - 8x + 6y + 9 = 0$

   b $(3, -4), 4$ *c* $(4, -3), 4$

10*a* $y^2 = -2x$ *b* $x^2 - 2xy + y^2 = -2\sqrt{2}(x + y)$; isometric transformation

11   $3y = 2x^2 + 9$ 12 $xy = 2$

13a $x^2 + y^2 = 4$ b $x^2 + 4y^2 = 4$

Page 137   Exercise 9B

1   $(3, 1), (4, 1), (5, 2), (4, 2)$; parallelogram, same area

2   $(1, 0), (4, 0), (6, 2), (3, 2)$; parallelogram, same area

3   $(7, 2), (9, 2), (12, 4), (10, 4)$; parallelogram, area four times as great

4   $(2, 0), (8, 0), (14, 4), (8, 4)$; parallelogram, area four times as great

5   $(2, 5), (3, 7), (3, 8), (2, 6)$; shear parallel to y-axis, same area
    $(1, 2), (4, 8), (4, 10), (1, 4)$; shear parallel to y-axis, same area

6   $\begin{pmatrix} 3 & 0 \\ 0 & 1 \end{pmatrix}$; $(0, 0), (3, 0), (3, 1), (0, 1)$; stretch in direction of x-axis

7 a $\{(4, 1), (0, 0), (-8, -2), (4a, a)\}$; points lie on $4y = x$

   b $\{(5, 0), (0, 0), (2, 0), (a + b, 0)\}$; points lie on x-axis

   c $\{(0, 4), (0, 0), (0, -8), (0, 4a)\}$; points lie on y-axis

   d $\{(0, 0)\}$; plane maps to origin

8   no inverse matrix; not a one-to-one correspondence of points

9 a $(0, 0), (2, 6), (6, 8), (4, 2)$ b $(0, 0), (4, 8), (6, 12), (2, 4)$

10   $(4, 2), (14, 6), (19, 9), (9, 5)$

Page 139   Exercise 10B

1   1, 4, 4   2   6, 6, 1   3   6, 24, -4 4   4, 100, 25

6   $|ad - bc| = 0$. Images lie on a straight line.

Geometry – Answers to Chapter 3
Page 142   Exercise 1

1   T   2   –   3   F   4   T   5   –   6   F   7   T   8   –

9   F   10   –   11   T   12   T   13   F   14   T   15a   F   b   T

   c   F   16   F   17   T   18   T   19   T   20   –

Page 144   Exercise 2

1   greater than 100   2 -1 3 $(-3, -5)$ 4 $\pm\sqrt{\dfrac{A}{\pi}}$ 5   negative

6   $\frac{1}{4}\pi$ 7   4   8 $\sqrt{[(x_2 - x_1)^2 + (y_2 - y_1)^2]}$ 9   T   10   F   11   T

12 T   13 F   14 T   15 F   16 F   17 T   18 F

19 $8 = 2^3 \Rightarrow \log_2 8 = 3$ 20 $f(x) = x^2 - \dfrac{1}{x^2} \Rightarrow f(-2) = 3\frac{3}{4}$

21 $y = \sin^2 x \Rightarrow \dfrac{dy}{dx} = 2\sin x \cos x$ 22 $x = \frac{3}{2}\pi \Rightarrow \sin x = -1$

Page 146   Exercise 3

1   If a number is divisible by 5, it ends in zero.

2   If a man is over 180 cm tall, he is a guardsman.

3   If n is an odd number, it is a prime number greater than 2.

4   If the diagonals of a quadrilateral intersect at right angles, the quadrilateral is a square.

5   If in $\triangle$ABC, B and C are both acute angles, then A is a right angle.

6   If $x^2 = 9$, $x = 3$.

7   If a plane figure has a centre of symmetry, it has two perpendicular axes of symmetry.

8   If $a + b$ is an even number, then a and b are odd numbers.

9   If k is a multiple of 3, then 3 is a root of $x^2 + x - k = 0$.

10   If PA $\neq$ PB, then P is on the same side of L as B.

12   If the areas of two triangles are equal, the triangles are congruent. T, F.

13   If two triangles are congruent three angles of one are respectively equal to three angles of the other. F, T.

14   If opposite angles of a quadrilateral are supplementary, the quadrilateral is cyclic. T, T.

15   If two angles are equal, they are in the same segment of a circle. T, F.

Page 147   Exercise 4

1   yes   2   yes   3   yes   4   no   5   no   6   yes

7   yes   8   no   9   yes   10   no

11   If two triangles have corresponding sides in proportion, the triangles are equiangular. True. Triangles are equiangular $\Leftrightarrow$ corresponding sides are in proportion.

12   If two chords of a circle are equidistant from the centre of the circle, the chords are equal. True. Chords are equidistant from centre of circle $\Leftrightarrow$ chords are equal.

13   If $\angle$BAC = $\angle$B'A'C', then AB$\parallel$A'B' and AC$\parallel$A'C'. False.

14   If PQSR is a parallelogram, then PQ is equal and parallel to RS. True. PQSR is a parallelogram $\Leftrightarrow$ PQ is equal and parallel to RS.

15 $A \subset B \Leftrightarrow A \cup B = B$ *17*   Diameter is perpendicular to chord $\Leftrightarrow$ diameter bisects chord.   *18*   Triangle has two axes of symmetry (and therefore three axes) $\Leftrightarrow$ triangle is equilateral.

Page 149   Exercise 5

1 $D \subset L$ *2* $P \cap E \neq \phi$ *3* $S \subset P$ *4* $R \subset P$ *5* $Q \subset R$

6 $P \cap E = \phi$ *7* $S \subset R$ *8*   If x is an integer with zero in the units place, then x is divisible by 5.

9 x is a surd $\Rightarrow x$ does not have a decimal equivalent.

10 x is an isosceles triangle $\Rightarrow$ the base angles of x are equal.

11a x is an even number greater than $2 \Rightarrow x$ is the sum of two prime numbers.

  b x is the sum of two prime numbers $\Rightarrow x$ is an even number greater than 2. $(2 + 3 = 5)$

12 x is a year divisible by $4 \Rightarrow x$ is a leap year. (2000)

Page 152   Exercise 6

1   (i) $C \subset T$
    (ii) $P \cap W \neq \phi$
    (iii) $P \cap E = \phi$.

2   *c*   *3*   *d*   *4*   For some real x, x^2 is not positive.

5   No pupils find mathematics difficult.   *6*   Some dogs like cats.

7   There is no positive integer x such that $x + 3 > 0$.

8   Some parallelograms do not have half turn symmetry. Statement is true.

9   Some schoolboys lie. Negation is true.

10   Some numbers which have zero in the units place are not divisible by five. Statement is true.

11   Some numbers of the form $2^n - 1$, n an integer, are not prime. Negation is true.

12   All matrices have a multiplicative inverse. Statement is true.

13   For some real numbers x, $x^2 + 1$ is not positive. Statement is true.

14   For some such pairs, $\overrightarrow{AB} \neq \overrightarrow{A_1 B_1}$. Statement is true.

15   For some such lines, DE is not parallel to BC. Statement is true.

Page 155   Exercise 7

1   valid   *2*   valid   *3*   invalid   *4*   valid   *5*   invalid

6   valid   *7*   valid   *8*   invalid   *9 a*   54 is divisible by 3; valid

  b   no valid conclusion   *10*   If mathematics is a useful subject people should study mathematics.

11 $x = 1$ or $x = 2$ *12* All squares can be inscribed in circles.

13a valid *b* invalid *c* invalid   *15a* valid *b* invalid

Page 159   Exercise 8

1 $y, z, 180$ *2* $2x = z = 2y$ *5* square, and add .

Geometry – Answers to Revision Exercises

Page 163   Revision Exercise 1

1 a $x^2 + y^2 = 144$ *b* $x^2 + y^2 = k^2$ *c* $x^2 + y^2 = 100$ *d* $x^2 + y^2 = 17$

2 $(\sqrt{5}, 2\sqrt{5}), (-\sqrt{5}, -2\sqrt{5})$ *3* $x^2 + y^2 = 16$; circle, centre O, radius 4

4   interior of circle, centre $(1, -1)$, radius 3   *5 a* $x^2 + y^2 - 4x - 6y - 23 = 0$

 b $x^2 + y^2 - 4x - 6y = 0$ *c* $x^2 + y^2 - 4x - 6y - 13 = 0$ *d* $x^2 + y^2 - 4x - 6y + 4 = 0$

 e $x^2 + y^2 - 4x - 6y + 9 = 0$ *6* $(3, 1), 10$; $x^2 + y^2 - 6x - 2y = 0$

7 a $(3, -5), 6$ *b* $(-1, 2), \sqrt{4\frac{2}{3}}$ *8 a* $x^2 + y^2 + 4x - 8y - 5 = 0$

 b $x^2 + y^2 - 4x - 8y - 5 = 0$ *c* $x^2 + y^2 = 25$ *10* $4x^2 + 4y^2 + 8x + y - 60 = 0$

11 $(1, 15), (-4, -10)$; $x + 5y = 11$ *12* $x + 3y + 1 = 0$

13 $(4, 3), (-4, 3)$; $4x + 3y = 25, -4x + 3y = 25$; $(0, 8\frac{1}{3})$ *14b* 6

15a $(7, -3), 10$; $(-5, 6), 5$ *b* $(-1, 3), 4x - 3y + 13 = 0$

16 $\pm\sqrt{7}y = 3(x - 8)$ *17* $\sqrt{3}y = \pm x + 2\sqrt{3}$ *18* $(-9, 0), (-9, -12)$

20 $(-2, -4), \sqrt{10}$ *21* $x^2 + y^2 + 2x - 10y + 8 = 0$; $(2, 2)$

22   touches at $(3\cos\alpha, 3\sin\alpha)$ *23* $y = 0, y = \frac{4}{3}x$ *24* $2x + y = 5$; $1\frac{4}{5}$

25 $(a\sin\theta, a\cos\theta), a\sin\theta$; $x = y\tan 2\theta$; $y = 0$

Page 165   Revision Exercise 2

1 a $(2, 0), (4, 0), (4, -1)$ *b* $(-2, 0), (-4, 0), (-4, -1)$

2 $(0, 0), (2, 1), (1, 5), (-1, 4)$ *3* $(4, 0), (9, 1)$; $5y = x - 4$

4 $(-2, 0), (0, 2), (2, 0), (0, -2)$. A square maps onto itself under quarter turns about

   O.   *5* $(0, 0), (12, 4), (16, 16), (4, 12)$; $[O, \frac{1}{2}]$, $\begin{pmatrix} \frac{1}{2} & 0 \\ 0 & \frac{1}{2} \end{pmatrix}$

6 $\begin{pmatrix} -1 & 0 \\ 0 & 1 \end{pmatrix}$; $\begin{pmatrix} 1 & 0 \\ 0 & -1 \end{pmatrix}$; $-1, 0, 0, -1$

7   dilatation $[O, 4]$; $\begin{pmatrix} 4 & 0 \\ 0 & 4 \end{pmatrix}$ *8 a* $\begin{pmatrix} -1 & 0 \\ 0 & -1 \end{pmatrix}$, half turn about O

 b $\begin{pmatrix} 1 & 0 \\ 0 & 1 \end{pmatrix}$, identity   *c* $\begin{pmatrix} 1 & 0 \\ 0 & 1 \end{pmatrix}$, identity   *9* θ radians

10a rotation about O of α radians *b* Rotation of α radians followed by a rotation of α radians = rotation of 2α radians. $\begin{pmatrix} \cos n\alpha & -\sin n\alpha \\ \sin n\alpha & \cos n\alpha \end{pmatrix}$ *11* Yes

12 Yes; reflection in y-axis, reflection in x-axis

14b $\begin{pmatrix} 0 & -1 \\ 1 & 0 \end{pmatrix}$, anticlockwise quarter turn about O

15 a M, I, M *b* (1) I (2) M *16* $\begin{pmatrix} 0 & -1 \\ 1 & 0 \end{pmatrix}$, anticlockwise quarter turn about O

17 $D = \begin{pmatrix} 3 & 0 \\ 0 & 3 \end{pmatrix}$, $H = \begin{pmatrix} -1 & 0 \\ 0 & -1 \end{pmatrix}$ *18* $(0, 0), (1, 1), (0, 2), (-1, 1)$

19 $x = 7$ *20* $x + 3y = 8; (2, 2); \frac{1}{4}\pi$ radians *21* $5x^2 + 6xy + 2y^2 = 9$

22a $x^2 + 4y^2 = 100$ *b* $(10, 0), (-10, 0), (0, 5), (0, -5)$

23b M, M. The determinant of M is zero.

24 $(0, 0), (a, c), (a + b, c + d), (b, d)$; each $= ad - bc$

25 Equation of image is of first degree in x and y.

26 $x - 3y = 6, 2x - 5y = 9; (-3, -3)$

27a $X = \begin{pmatrix} -1 & 0 \\ 0 & -1 \end{pmatrix}$ *b* $y = x^2 + 5x + 4, y = -x^2 + 5x - 4, y = -x^2 - 5x - 4$

Page 170   Revision Exercise 3

1 a T *b* F *c* T *d* F. In *b*, $a \neq 0$; in *d*, $c > 0$.

2 a ± 19 *b* $\{x : 1 < x \leqslant 6\}$ *c* $-2 \leqslant x \leqslant 3$ *d* It has two perpendicular axes of symmetry, *or* it is a rectangle, *or* a rhombus. *3* No. $x = 0 \Rightarrow x(x - 3) = 0$; yes.

4 a Every relation is a mapping. *b* If a triangle has an angle of 80°, it has angles of 40° and 60°. *c* All real numbers are rational. *d* If $x \in R$ and $x^2 > x$, then $x > 1$.

5 $x, y \in \{\text{positive real numbers}\}$ *6 a* $\Rightarrow$ *b* $\Leftrightarrow$ *c* $\Leftrightarrow$ *d* $\Leftrightarrow$

7 C *8* 4 *9* 5. A has 32 subsets.

10a valid *b* invalid *11a* invalid *b* valid *13* Suppose that *a* is not parallel to *b*. Then *a* and *b* must meet at a point P. Hence through P there are two lines both perpendicular to *c*. This is false, since there is only one line from a point perpendicular to a line. Therefore *a* must be parallel to *b*.

Geometry – Answers to Cumulative Revision Section

Page 176   Exercise 1

1 a $\frac{2}{5}$ *b* -1 *c* not defined *d* -2 *e* 0 *f* $\dfrac{k - q}{h - p}$, $h \neq p$

2 $y = -\frac{1}{2}x - 3$ *3 a* $2x - 5y + 13 = 0$ *b* $x + y - 8 = 0$ *c* $x = 1$

4 $y = \frac{2}{3}x + \frac{4}{3}, y = \frac{2}{3}x - \frac{7}{6}$; equal gradients *5* $m_1 m_2 = -1$

6 *a* $m_1m_2 = -1$ *b* $AB^2 + BC^2 = AC^2$ 7 *a* $4x + 3y = 37$ *b* $(4, 7)$

8     $(1, 6)$ 9 *a* $x + y - 11 = 0$ *b* $(7, 4), x + y - 11 = 0$ *c* Triangle is isosceles

10*a* . $x + 10y - 16 = 0$ *b* $(-2, -4), (-1, 6)$ *c* $10x - y + 16 = 0$

11*b* $m_{PQ} = -1$ *c* $(0, 7)$ 12*a* $2x + 3y = 4$ *b* $(2, 0), (0, \frac{4}{3})$ *c* $3:2$

14   $x - 2y + 7 = 0$ 15 $-3:1$

16*a* $2x + 3y = 13$ *b* linear equation   *c* same equation

17   $P(p\cos\alpha, p\sin\alpha)$ 18 $(3, 1), (-1, 5)$ and $(-3\frac{2}{5}, 7\frac{2}{5})$

Page 178   Exercise 2

1 *a* $\begin{pmatrix} 4 \\ 6 \\ -2 \end{pmatrix}$ *b* $\begin{pmatrix} 2 \\ 7 \\ -5 \end{pmatrix}$ *c* $\begin{pmatrix} -2 \\ -7 \\ 5 \end{pmatrix}$ *d* $\begin{pmatrix} -2 \\ 3 \\ 1 \end{pmatrix}$ *e* $\begin{pmatrix} 2 \\ -3 \\ -1 \end{pmatrix}$ 2 $h = -2, k = -6; 1:2$

3 *a* $-i + 5j + k$ 4 *a* 7 *b* $(1, 1\frac{2}{3}, 0)$

5 *a* $a = i + 4j - 2k, b = 6i - 2j - k, c = 12i - 4j - 2k$ *b* $a.b = 0$ *c* $c = 2b$

6 *a* $2v$ *b* $v - u$ *c* $-u$ *d* $-v$ *e* $u - v$ *f* $-v$ *g* $u - v$ *h* u *i* v

   j $v - u$ *k* $-u$; 12·5

7 *a* $\overrightarrow{PQ} = 3\overrightarrow{QR}$ *b* $3:1$ *c* $-4:3$ 8 *a* O, A, B, C collinear *b* $k = 7$

9 *a* (1) 1 (2) 1·5 (3) 3 (4) 1 *b* 25, 5 *c* 3·5   10*b* $(r - p).(s - q) = 0$

11    $p - q + r$; parallelogram

13    $a_1{}^2 + a_2{}^2 + a_3{}^2 = 1, 3a_1 - 2a_2 - a_3 = 0, a_1 + 3a_2 - 4a_3 = 0; \pm 1/\sqrt{3}$ for each

14*a* F *b* F *c* F *d* T

Geometry – Answers to Cumulative Revision Exercises

Page 181   Cumulative Revision Exercise 1

1

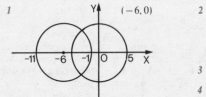

2 *a* $y = x + 8$

   b $\{(x, y): x^2 + y^2 \leqslant 64, y \geqslant x + 8\}$

   c $16(\pi - 2)$ square units

3    $v + u, v - u, v - 2u, u - 2v$

4   N is midpoint of AC

5 *a* opposite angles supplementary   *b* draw circle with diameter OP

Page 182   Cumulative Revision Exercise 2

1    $\frac{3}{4}u, \frac{1}{2}v + \frac{1}{4}u$ 2 *a* $\begin{pmatrix} -1 \\ 1 \end{pmatrix}, \begin{pmatrix} 3 \\ 4 \end{pmatrix}$ *b* $\begin{pmatrix} 4 \\ 3 \end{pmatrix}$ *c* 5   3 *a* (20, 14) *b* $(-8, -10)$

4 *a* $2y = x, \frac{1}{2}$ *b* (1) $(4, -2)$ (2) $2y = -x$ *c* translation $\begin{pmatrix} 0 \\ 4 \end{pmatrix}$; $2y = x + 8$

5 $\begin{pmatrix} 1 & 1 \\ 1 & -1 \end{pmatrix}$ *a* line $y = x$ *b* line $y = -x$ 6 47·2°   7 21°

8 four pairs of angles in the same segment, etc.   9 *b* dilatation $[O, \sqrt{2}]$

Page 183 Cumulative Revision Exercise 3

1 *a* $2x - 3y + 13 = 0, 3x + 2y = 0$ *b* B(1, 5), C(4, −6) *c* M($2\frac{1}{2}, -\frac{1}{2}$), D(7, −4)

2 *a* (4, −3), 5; no constant term, hence (0, 0) satisfies equations *b* (8, 0)

 c $4x - 3y = 0, 4x + 3y = 32$ *d* (4, $5\frac{1}{3}$)

3 *a* (2, 7, −5) *b* $\begin{pmatrix} 8 \\ -8 \\ 4 \end{pmatrix}, \begin{pmatrix} 5 \\ 1 \\ -2 \end{pmatrix}, \begin{pmatrix} 0 \\ 16 \\ -12 \end{pmatrix}$ *c* cosine of each angle = $2/\sqrt{30}$

4 *a* identity; rotation of $\frac{1}{2}\pi$ radians about O; rotation of $-\frac{1}{2}\pi$ radians about O; half turn about O

b

×	I	P	Q	H
I	I	P	Q	H
P	P	H	I	Q
Q	Q	I	H	P
H	H	Q	P	I

c identity

reflection in *x*-axis
reflection in *y*-axis

half turn about O

×	I	R	S	H
I	I	R	S	H
R	R	I	H	S
S	S	H	I	R
H	H	S	R	I

d In the second table each element is its own inverse.

Page 184 Cumulative Revision Exercise 4

1 *b* A(1, 2), B(−1, −1), C(7, −2)   2 *a* $x^2 + y^2 - 4x - 6y - 5 = 0$ *d* $7/\sqrt{2}$

3 *a* (1) $|a||b|\cos\theta$ (2) $a_1b_1 + a_2b_2 + a_3b_3$; $\cos\theta = \dfrac{a_1b_1 + a_2b_2 + a_3b_3}{\sqrt{(a_1^2 + a_2^2 + a_3^2)(b_1^2 + b_2^2 + b_3^2)}}$

 b $\begin{pmatrix} -2 \\ -1 \\ 2 \end{pmatrix}, \begin{pmatrix} -6 \\ 3 \\ -2 \end{pmatrix}$ *c* 76·2° or 76·3°; 10·2

4 *a* (1) (−2, −2), (−4, −1), (−4, −2)   (2) (2, 2), (4, 1), (4, 2)

 b half turn about O   (1) $x = 0$ (2) $y = -x$ (3) $y = -\frac{1}{2}x$ (4) $y = -\dfrac{1}{m}x$

5 *a* $y = 2x - 3$ *b* $x + 2y + 3 = 0$ *c* $3x + y + 3\sqrt{2} = 0$

Page 185 Cumulative Revision Exercise 5

1 *a* P(3, 2), Q($5\frac{1}{2}, 3\frac{1}{4}$), R(7, 4); $m_{PQ} = m_{QR}$ *b* 5:3

2 *a* (0, −3), (2, 1) *c* $k = -\frac{3}{2}$; ($\frac{3}{2}, -\frac{5}{4}$), $\frac{1}{4}\sqrt{85}$

3 *b* $\boldsymbol{u} \cdot \boldsymbol{v} = 0$ *c* $-\frac{1}{3}, \frac{2}{3}$ and $\frac{2}{3}$; or $\frac{1}{3}, -\frac{2}{3}, -\frac{2}{3}$ *d* $p = 1, q = 2, r = 3$

4 *a* A$_1$(10, 0), B$_1$(20, 0), C$_1$(20, 10), D$_1$(10, 10) *b* A$_2$(8, 6), B$_2$(16, 12), C$_2$(10, 20), D$_2$(2, 14) *c* $\begin{pmatrix} 5 & 0 \\ 0 & 5 \end{pmatrix} \begin{pmatrix} \frac{4}{5} & -\frac{3}{5} \\ \frac{3}{5} & \frac{4}{5} \end{pmatrix} = \begin{pmatrix} \frac{4}{5} & -\frac{3}{5} \\ \frac{3}{5} & \frac{4}{5} \end{pmatrix} \begin{pmatrix} 5 & 0 \\ 0 & 5 \end{pmatrix}$

 d dilatation; rotation about O through θ radians where $\cos\theta = \frac{4}{5}, \sin\theta = \frac{3}{5}$

5 *a* equiangular *b* 2 *c* 120°

Page 186   Cumulative Revision Exercise 6

1 $x - y = 3,\ x + 3y = -1;\ (2, -1);\ 3x + y = 5$

2 a $(6, 1);$ *2 b* $y = 1$ *3 a* $T = \begin{pmatrix} 3 & 2 \\ 4 & 3 \end{pmatrix},\ T^{-1} = \begin{pmatrix} 3 & -2 \\ -4 & 3 \end{pmatrix};\ 5x - 4y + 3 = 0$

 b $\begin{pmatrix} 3 & 2 \\ 4 & 3 \end{pmatrix}\begin{pmatrix} -10 & 15 \\ 15 & -20 \end{pmatrix} = \begin{pmatrix} 0 & 1 \\ 1 & 0 \end{pmatrix}\begin{pmatrix} 5 & 0 \\ 0 & 5 \end{pmatrix}$ *4 a* $\mathbf{k} = \frac{1}{3}(\mathbf{a} + \mathbf{b} + \mathbf{c})$, etc

 b $\overline{\text{KL}} = \overline{\text{NM}}$ *5 a* $x^2 + y^2 - 4x - 2y - 20 = 0$ *b* $(2, 1),\ 5$

6 a $109 \cdot 5°$ *b* $(1, 0, -3)$

Trigonometry – Answers to Chapter 1

Page 190   Exercise 1

1 $\cos(A + B) + \cos(A - B)$ *2* $\cos(x + y) + \cos(x - y)$ *3* $\cos(p + q) + \cos(p - q)$

4 $\cos 80° + \cos 20°$ *5* $\cos 50° + \cos 20°$ *6* $\cos 66° + \cos 40°$

7 $\cos(A - B) - \cos(A + B)$ *8* $\cos(x - y) - \cos(x + y)$ *9* $\cos(p - q) - \cos(p + q)$

10 $\cos 40° - \cos 80°$ *11* $\cos 15° - \cos 35°$ *12* $\cos 20° - \cos 50°$

13 $\cos 2x + \cos 2y$ *14* $\cos 2y - \cos 4x$ *15* $\cos \alpha + \cos \beta$

16 $\cos(2B - 2C) - \cos 2A$ *17* $\cos 2\theta + 1$ *18* $-\cos 2\theta$

19 $\frac{1}{2}(\cos 80° + \cos 20°)$ *20* $\frac{1}{2}(\cos 88° + \cos 44°)$ *21* $\frac{1}{2}(\cos 110° - \cos 150°)$

22 $\cos 220° - 1$ *23* $\frac{1}{2}$ *24* -1

Page 191   Exercise 2

1 $\sin(A + B) + \sin(A - B)$ *2* $\sin(x + y) + \sin(x - y)$ *3* $\sin(p + q) + \sin(p - q)$

4 $\sin 80° + \sin 20°$ *5* $\sin 50° + \sin 20°$ *6* $\sin 84° + \sin 36°$

7 $\sin(A + B) - \sin(A - B)$ *8* $\sin(x + y) - \sin(x - y)$ *9* $\sin(p + q) - \sin(p - q)$

10 $\sin 80° - \sin 40°$ *11* $\sin 80° - \sin 70°$ *12* $\sin 100° + \sin 50°$

13 $\frac{1}{2}(\sin 6\theta + \sin 4\theta)$ *14* $\frac{1}{2}(\sin 5\theta - \sin \theta)$ *15* $\frac{1}{2}(\sin 2\theta + \sin \theta)$

16 $\sin 2P + \sin 2Q$ *17* $\sin 2\theta$ *18* $\sin \alpha + \sin \beta$

19 $1 - \sin 2\alpha$

Page 192   Exercise 3B

1 a $1, -1$ *b* $1, -1$ *c* $2, -2$ *d* $3, -3$ *e* $1, -1$

2 $\cos 2x° + \cos 90° = \cos 2x°;\ 1, -1$ *3 a* $\cos 2x° + \cos 60° = \cos 2x° + \frac{1}{2};$

      $1 \cdot 5,\ -0 \cdot 5$ *b* $\cos \frac{3}{2}\pi - \cos 2\theta = -\cos 2\theta;\ 1, -1$ *4* -1

9 $\frac{1}{2}(\cos 2\alpha - \cos 4\alpha);\ \frac{1}{2}(\cos 2\alpha - \cos 128\alpha)$ *10* $\frac{1}{2}(\sin 128\alpha - \sin 2\alpha)$

Page 194   *Exercise 4*

1 $2\cos\frac{1}{2}(A+B)\cos\frac{1}{2}(A-B)$ *2* $2\cos 2X\cos X$ *3* $2\cos 25°\cos 15°$

4 $2\cos\frac{1}{2}(P+Q)\cos\frac{1}{2}(P-Q)$ *5* $2\cos 3Y\cos 2Y$ *6* $\cos 20°$

7 $-2\sin\frac{1}{2}(A+B)\sin\frac{1}{2}(A-B)$ *8* $-2\sin 3X\sin X$ *9* $-2\sin 35°\sin 15°$

10 $-2\sin\frac{1}{2}(P+Q)\sin\frac{1}{2}(P-Q)$ *11* $2\sin 2Y\sin Y$ *12* $-\sin 10°$

13 $2\sin\frac{1}{2}(A+B)\cos\frac{1}{2}(A-B)$ *14* $2\sin 3X\cos 2X$ *15* $2\sin 20°\cos 5°$

16 $2\sin\frac{1}{2}(P+Q)\cos\frac{1}{2}(P-Q)$ *17* $2\sin 2Y\cos Y$ *18* $2\cos 80°$

19 $2\cos\frac{1}{2}(A+B)\sin\frac{1}{2}(A-B)$ *20* $2\cos 2X\sin X$ *21* $2\cos 33°\sin 11°$

22 $2\cos\frac{1}{2}(P+Q)\sin\frac{1}{2}(P-Q)$ *23* $-2\cos 3Y\sin Y$ *24* 0

25 0 *26* $\cos 12°$ *27* $-2\sin 110°$ or $-2\sin 70°$

28 $2\sin 2\alpha\cos\beta$ *29* $2\cos\alpha\cos\beta$ *30* $2\cos(x+\frac{1}{2}h)\sin\frac{1}{2}h$

31 $-2\sin(x+\frac{1}{2}h)\sin\frac{1}{2}h$ *32* $2\cos\theta$ *33* $\sin 2\theta$

36b ϕ

Page 196   *Exercise 5*

1 $\{60, 300\}$ *2* $\{90\}$ *3* $\{45, 225\}$ *4* $\{135, 225\}$ *5* $\{30, 210\}$

6 $\{210, 330\}$ *7* $\{30, 60, 210, 240\}$ *8* $\{30, 150, 210, 330\}$

9 $\{30, 120, 210, 300\}$ *10* $\{30, 150, 270\}$ *11* $\{60, 180, 300\}$

12 $\{0, 60, 120, 180, 240, 300\}$ *13* $\{\frac{1}{4}\pi, \frac{3}{4}\pi\}$ *14* $\{0, \pi\}$ *15* $\{\frac{2}{3}\pi, \frac{4}{3}\pi\}$

16 $\{0 < x < 180\}$ *17* $\{90 < x < 270\}$ *18* $\{30 \leqslant x \leqslant 150\}$

Page 196   *Exercise 5B*

1 $\{60+n.360\} \cup \{300+n.360\}$ *2* $\{90+n.360\}$

3 $\{45+n.360\} \cup \{225+n.360\}$ *4* $\{135+n.360\} \cup \{225+n.360\}$

5 $\{30+n.360\} \cup \{210+n.360\}$ *6* $\{210+n.360\} \cup \{330+n.360\}$

Page 197   *Exercise 6*

1 $\{0, 90, 180, 270\}$ *2* $\{45, 90, 135, 225, 270, 315\}$

3 $\{0, 45, 135, 180, 225, 315\}$ *4* $\{0, 90, 180, 270\}$ *5* $\{45, 120, 135, 225, 240, 315\}$

6 $\{0, 60, 90, 180, 270, 300\}$ *7* $\{0, 90, 180, 210, 270, 330\}$

8 $\{30, 45, 135, 150, 225, 315\}$ *9* $\{0, 60, 90, 120, 180, 240, 270, 300\}$

10 $\{30, 90, 150, 210, 270, 330\}$ *11* $\{0, \frac{1}{6}\pi, \frac{1}{2}\pi, \frac{5}{6}\pi, \pi, \frac{7}{6}\pi, \frac{3}{2}\pi, \frac{11}{6}\pi\}$

12 $\{0, \frac{1}{3}\pi, \frac{2}{3}\pi, \pi, \frac{4}{3}\pi, \frac{5}{3}\pi\}$ *13* $\{\frac{1}{3}\pi, \pi, \frac{5}{3}\pi\}$ *14* $\{0, \frac{1}{3}\pi, \pi, \frac{5}{3}\pi\}$

Page 197   Exercise 7

1   {30, 150}   *2*   {70·5, 180, 289·5}   *3*   {0, 120, 240}   *4*   {30, 90, 150, 270}

5   {0, 60, 300}   *6*   {0, 180}   *7*   {90}   *8*   {60, 126·9, 233·1, 300}

9   {90, 236·4, 270, 303·6}   *10*   {0, 90, 180}   *11*   {0, 48·2, 311·8}   *12* ∅

13   {$\frac{2}{3}\pi, \pi, \frac{4}{3}\pi$}   *14*   {$\frac{1}{6}\pi, \frac{5}{6}\pi, \frac{3}{2}\pi$}

Trigonometry – Answers to Chapter 2

Page 203   Exercise 1

1 a $\sqrt{2}, 45$ *b* $\sqrt{2}, 135$ *c* 2, 60   *d* 2, 330   *e* 5, 53·1   *f* 13, 112·6

3 a $5\cos(x-36·9)°$ *b* $\sqrt{2}\cos(x-225)°$ *c* $\sqrt{10}\cos(x-288·4)°$

4 a $\sqrt{2}\cos(\theta-\frac{1}{4}\pi)$ *b* $2\cos(\theta-\frac{2}{3}\pi)$ *5 a* $10\cos(x-36·9)°$ *b* $\sqrt{10}\cos(x-341·6)°$

  c $2\cos(x-60)°$ *d* $\sqrt{2}\cos(x-135)°$ *e* $17\cos(x-61·9)°$ *f* $2\sqrt{2}\cos(x-225)°$

6 $5\cos\omega t+2\sqrt{3}\sin\omega t; \sqrt{37}\cos(\omega t-0·61)$

Page 204   Exercise 2

1   53·1   *2*   292·6   *3*   108·2, 348·2   *4*   333·4(5)   *5*   180, 270

6   114·3, 335·7   *7*   0, 240   *8*   167·3, 347·3

9 a $2\sin(x-60)°$; 90, 210   *b* $10\sin(x-53·1)°$; 58·8, 227·4   *10a* $\frac{1}{2}\pi, \frac{11}{6}\pi$

  b $\frac{1}{2}\pi, \pi$ *11* $5\cos(2x-53·1)°$; 26·6   *12* $10\cos(2x-36·9)°$; 48·4, 168·4

Page 207   Exercise 3

1 a Max. 1 at $x=0$; min. -1 at $x=180$ *b* Max. 2 at $x=30$; min. -2 at $=210$

  c Max. 5 at $x=200$; min. -5 at $x=20$

2 a $5\cos(x-53·1)°$ *b* Max. 5 at $x=53·1$; min. -5 at $x=233·1$ *c* $x=143·1, 323·1$

3 a $2\cos(x-60)°$ *b* Max. 2 at $x=60$; min. -2 at $x=240$ *c* $x=150, 330$

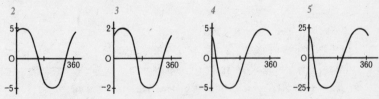

4 a $5\cos(x-306·9)°$ *b* Max. 5 at $x=306·9$; min. -5 at $x=126·9$

  c $x=36·9, 216·9$ *5 a* $25\cos(x-286·3)°$

  b Max. 25 at $x=286·3$; min. -25 at $x=106·3$ *c* $x=16·3, 196·3$

6   7   8 *c*   10 $38\sqrt{5}\cos(30t-63\cdot4)^\circ$.
H.W. at 02 07 (2·11) and
14 07 hours; L.W. at 08 07
and 20 07 hours. 00 19,
03 55, 12 19, 15 55 hours.

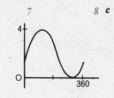

11b $\sqrt{13}\cos(a-56\cdot3)^\circ$; $3\cos a^\circ-2\sin a^\circ=\sqrt{13}\cos(a-326\cdot3)^\circ$

c $\sqrt{13}$ in each case   *12* $38+30\cos\theta+20\sin\theta$; 8·61 m

Trigonometry — Answers to Revision Exercises

Page 210   Revision Exercise 1

1 a $\cos(X+Y)+\cos(X-Y)$ *b* $\cos80^\circ+\cos20^\circ$ *c* $\cos(X-Y)-\cos(X+Y)$

d $\cos\alpha+\cos\beta$ *e* $\cos q-\cos p$ *2 a* $\sin(X+Y)+\sin(X-Y)$

b $\sin70^\circ+\sin10^\circ$ *c* $\sin(R+S)-\sin(R-S)$ *d* $\sin\alpha-\sin\beta$ *e* $\sin2x+\sin2\theta$

3 a $\frac{1}{2}$ *b* $\frac{1}{2}\sqrt{3}+\sin20^\circ$ *c* $\sin50^\circ$ *d* $\cos\frac{1}{2}x^\circ-\cos x^\circ$

e $\frac{1}{2}(\cos3a^\circ+\cos a^\circ)$ *f* $\frac{1}{2}(\cos2b^\circ-\cos6b^\circ)$

4 a $\cos2\theta+\cos4\theta$ *b* $\frac{1}{2}+\frac{1}{2}\sin2a^\circ$ *c* $\sin2\theta-\frac{1}{2}\sqrt{3}$ *d* $\frac{1}{2}-\sin2x^\circ$ ·

5 a $1,-1$ *b* $5,-5$ *c* $1,-1$ *d* $1,-1$ *e* $1,-1$

6 a $\cos(2x+10)^\circ+\cos30^\circ$; $1+\frac{1}{2}\sqrt{3}$, $-1+\frac{1}{2}\sqrt{3}$ *b* $\cos2\theta-\cos\frac{1}{3}\pi$; $\frac{1}{2}$, $-1\frac{1}{2}$

7 a $2\cos40^\circ\cos20^\circ$ *b* $2\sin70^\circ\cos10^\circ$ *c* $-2\cos15^\circ\sin5^\circ$

d $2\sin40^\circ$ *e* $2\sin x^\circ\cos\frac{1}{4}x^\circ$ *f* $-2\sin5\theta\sin2\theta$

8 a $2\cos(\theta+\frac{1}{2}h)\sin\frac{1}{2}h$ *b* $-2\sin(\theta+\frac{1}{2}h)\sin\frac{1}{2}h$ *c* $\sqrt{3}\cos\alpha$ *d* 0

9 a $\sqrt{3}$ *b* $-\tan\theta$ *10* $x=2\sin3\theta\cos\theta$, $y=2\cos3\theta\cos\theta$

12a $\{60,240\}$ *b* $\{90,270\}$ *c* $\{75,105,195,225,315,345\}$

13a $\{30,90,150,210,270,330\}$ *b* $\{0,45,135,225,315\}$

c $\{0,30,60,120,150,180,210,240,300,330\}$ *d* $\{0,15,105,135,180,225,255,345\}$

14a $\{90,210,270,330\}$ *b* $\{30,150,270\}$ *c* $\{90\}$ *d* $\{60,180,300\}$

17a   maximum $\frac{3}{4}$, when $x=40$; minimum $-\frac{1}{4}$, when $x=130$.

b·   maximum $\frac{1}{2}$, when $x=165$; minimum $-\frac{1}{2}$, when $x=75$.

19a $\{45,120,135,225,240,315\}$ *b* $\{0,60,120,180,240,300\}$

21a $2\sin(\alpha+\beta)\sin(\alpha-\beta)=\cos2\beta-\cos2\alpha$

Page 212   Revision Exercise 2

1 *a* $\sqrt{2}$, 45   *b* $\sqrt{2}$, 315   *c* 2, 30   *d* 2, 150   *e* 10, 36·9   *f* 5, 216·9

3 *a* $\sqrt{74}\cos(x-324\cdot5)°$ *b* $\sqrt{13}\cos(x-213\cdot7)°$

4 *a* $4\sqrt{2}\cos(x-\frac{7}{4}\pi)$ *b* $2\cos(x-\frac{11}{6}\pi)$ 5 *a* $\sqrt{85}\cos(x-310\cdot6)°$

  b $\sqrt{41}\cos(x-51\cdot3)°$ *c* $\sqrt{34}\cos(x-149)°$ *d* $\sqrt{85}\cos(x-12\cdot5)°$

6 *a* $\sqrt{85}$, $-\sqrt{85}$ *b* $\sqrt{41}$, $-\sqrt{41}$ *c* $\sqrt{34}$, $-\sqrt{34}$ *d* $\sqrt{85}$, $-\sqrt{85}$

7   2, -2; 150, 330; 60, 240

8   10, -10; 53·1, 233·1; 143·1, 323·1

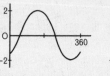
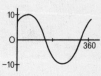

9   1, -25 10 $\frac{1}{2}, \frac{1}{6}$ 11 $\{90, 330\}$ 12 $\{26\cdot6, 206\cdot6\}$

13*a* ϕ *b* $\{-\frac{1}{4}\pi\}$ *c* $\{\frac{11}{6}\pi\}$ *d* ϕ 14*a* 17 cm   *b* 1·33 s

Trigonometry – Answers to Cumulative Revision Section

Page 218   Exercise 1

1 *a* F   *b* T   *c* T   2 *a* (1·80, 2·40)   *b* (4·75, $-1\cdot56$)

3 *a* $\frac{12}{13}, \frac{12}{5}$ *b* $\frac{9}{41}, -\frac{40}{41}$ 4 *a* $\frac{1}{2}$ *b* $1/\sqrt{2}$ *c* $-1/\sqrt{2}$ *d* $\sqrt{3}$ *e* $-\sqrt{3}$

5 *a* $\{210, 330\}$ *b* $\{30, 330\}$ 6 45   7 6·49 cm²   8 29°

9   10·6 cm, 13·0 cm; 66·5 cm²   10 51·7°, 128·3°   11*a* 4·76 m   *b* 8·34 m²

12*a* 38·2 cm²   *b* 10·7 cm   *c* 58·3°, 48·7°   13 122 km, 076°

14*a* 12·6 cm, 13·6 cm   *b* 38·7°, 17·1°   *c* 38·7°   15*a* (5, 0, 3), (0, 0, 3), (0, 4, 3)

  b 36·9°   *c* 25·1°   *d* (2·5, 2, 1·5)   *e* 90°   *f* 53·1°

16*a* 1·89 m   *b* 67·1°   17*a* 39·0°   *b* 69·3°

Page 220   Exercise 2

1 *a* $-0\cdot342$ *b* $-0\cdot940$ *c* 0·364   *d* 2·824

2 *a* k *b* $-k$ *c* $-k$ *d* $-k$ 3 *a* $\frac{1}{6}\pi, \frac{1}{4}\pi, \frac{2}{3}\pi, \frac{4}{3}\pi, \frac{5}{3}\pi, 2\pi$

  b 60°, 90°, 270°, 150°, 105°   4 *a* 144°   *b* 2·51 radians   5 25·1 cm

6 *a* $\{30, 150, 390, 510\}$ *b* $\{45, 225, 405, 585\}$ *c* $\{135, 225, 495, 585\}$

8 $\frac{33}{65}, \frac{7}{25}$ 9 *a* -1 *b* 1   10*a* $\frac{1}{2}(1+\cos 2\theta)$ *b* $\frac{1}{2}(1+\cos x)$

11*a* 1   *b* $3\sin 2x$ *c* 0   *d* $\cos 8x$ 12*a* $\{90, 120, 240, 270\}$

  b $\{0, 30, 150, 180, 360\}$ *c* $\{0, 60, 180, 300, 360\}$ *d* $\{0, 30, 150, 180, 210, 330, 360\}$

13 $\frac{33}{65}$ 14 $-\frac{56}{65}$ 15*a* T   *b* T   *c* F   *d* T

18*b* $\sin 3\theta = 3\sin\theta - 4\sin^3\theta$ 19 $kt^2 + 2t - k = 0$

Trigonometry – Answers to Cumulative Revision Exercises

Page 223   Cumulative Revision Exercise 1

1 a $\frac{4}{5}, \frac{4}{3}$ *b* $\frac{12}{13}, -\frac{5}{13}$ *2 a* $-\sin\theta$ *b* $\cos\theta$ *c* $\sin\theta$ *d* $-\cos\theta$

     e $\cos\theta$ *f* $\cos\theta$ *3* at $(0,0)$ *4* $\frac{63}{65}, \frac{33}{56}$

6 $\{41{\cdot}4, 180, 318{\cdot}6\}$ *7 a* $\sin\alpha + \sin\beta$ *b* $\sqrt{3}/2 - \cos 2x$

8 a $2\sin 5\theta \sin 2\theta$ *b* $2\cos(x+k)\sin k$ *c* $\sqrt{3}\cos 10°$ *9* $\{0, 90, 180, 270, 360\}$

10a $x = 2\sin 2\theta \cos\theta, y = 2\cos 2\theta \cos\theta$ *11a* $2, -2$ *b* $1, 0$ *c* $1, -1$ *d* $5, -1$

12a $5\cos(x-143{\cdot}1)°$ *b* max. 5 at $x = 143{\cdot}1$, min. -5 at $x = 323{\cdot}1$ *13* $\{28{\cdot}1\}$

Page 224   Cumulative Revision Exercise 2

1 $\sin 30°, \sin\frac{1}{4}\pi, \sin 50°, \sin\frac{1}{3}\pi, \sin 70°, \sin\frac{1}{2}\pi$ *3* $-\frac{24}{25}, -\frac{16}{65}$

4 $\frac{1}{2}$ *5* $\{270\}$ *6 a* $\cos 2\theta + \cos 6\theta$ *b* $\frac{1}{2}(1 - \sin 40°)$

7 a $\tan 3x°$ *b* $\{45, 105, 165, 225, 285, 345\}$ *8* $2\cos 2\theta \sin\theta$;

     $\{\frac{1}{6}\pi, \frac{1}{4}\pi, \frac{3}{4}\pi, \frac{5}{6}\pi, \frac{5}{4}\pi, \frac{3}{2}\pi, \frac{7}{4}\pi\}$ *10a* $2\cos(x-60)°$ *b* max. 2 at $x = 60$;

     min. -2 at $x = 240$ *11a* $\{20{\cdot}6, 306{\cdot}8\}$ *b* $\{72{\cdot}4, 220{\cdot}2\}$

12a $d = b\cos\theta + a\sin\theta$ *b* $\begin{pmatrix} \cos\theta & -\sin\theta \\ \sin\theta & \cos\theta \end{pmatrix}$; rotations about O of θ and $\frac{1}{2}\pi$ radians

Calculus – Answers to Chapter 1

Page 229   Exercise 1

1 a $\cos x$ *b* $4\cos x$ *c* $\cos x$ *d* $1 - 2\cos x$

2 a $-\sin x$ *b* $-3\sin x$ *c* $-\sin x$ *d* $2x + \sin x$

3 a $2\cos x$ *b* $-5\sin x$ *c* $\cos x - \sin x$

     d $3x^2 + \cos x$ *e* $-\sin x - 1$ *f* $-2\sin x + 3\cos x$

4 a 1 *b* 0 *c* 1 *d* -1 *e* $\frac{1}{2}$ *f* $1/\sqrt{2}$ *g* $\frac{1}{2}$ *h* 0 *i* $\sqrt{3}/2$ *j* $-\sqrt{3}/2$

5 a 1 *b* 0 *c* -1 *d* -1

7 a $\sqrt{2}y - 1 = x - \frac{1}{4}\pi$ *b* $2y - \sqrt{3} = -x + \frac{1}{6}\pi$ *10b* 1

11a $(\frac{3}{4}\pi, 0), (\frac{7}{4}\pi, 0)$ *b*
      $(\frac{1}{4}\pi, 1{\cdot}4)$, maximum;
      $(\frac{3}{4}\pi, -1{\cdot}4)$, minimum

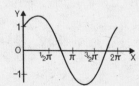

Page 231   Exercise 2

1 a $\sin x + C$ *b* $2\sin x + C$ *c* $x - \sin x + C$ *d* $\frac{1}{2}x^2 + 3\sin x + C$

2 a $-\cos x + C$ *b* $-4\cos x + C$ *c* $4x + \cos x + C$ *d* $-2\cos x + x^3 + C$

3 a $\sin x - \cos x + C$ *b* $3\sin x + 4\cos x + C$

4 a $\frac{1}{2}$ *b* 1 *c* 2   *5 a* $\frac{1}{2}\pi + 1$ *b* $\frac{1}{3}\pi + 1$ *c* 0

6   0   *7* 2   *8* $\frac{1}{6}\pi + \frac{1}{2}$ *9* $\sqrt{2} - 1$

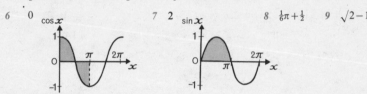

10 $\frac{1}{2}x + \frac{1}{2}\sin x + C$ *11* $\frac{1}{2}(1 - \cos x); \frac{1}{2}x - \frac{1}{2}\sin x + C$ *12a* $\frac{7}{24}\pi^3 + 5$ *b* -5

13 $1 - \frac{1}{4}\pi, 2 - \frac{1}{2}\pi, \frac{3}{4}\pi - 1$

Page 233   Exercise 3

$F(x) = f(g(x))$, where:

1 $f(x) = x^6, g(x) = x + 1$ *2* $f(x) = x^4, g(x) = 2x + 3$

3 $f(x) = \sqrt{x}, g(x) = x + 2$ *4* $f(x) = \sin x, g(x) = 3x + 4$

5 $f(x) = \cos x, g(x) = ax + b$ *6* $f(x) = x^2, g(x) = \sin x$

7 $f(x) = x^3, g(x) = x^2 + x + 1$ *8* $f(x) = \sqrt[3]{x}, g(x) = x^2 - 1$

9 $f(x) = x^3, g(x) = 1 + \cos x$ *10* $f(x) = \dfrac{1}{x}, g(x) = x + 3$

11 $f(x) = x^6, g(x) = x + \dfrac{1}{x}$ *12* $f(x) = \dfrac{1}{x}, g(x) = 5x - 4$

Page 236   Exercise 4

1 a $6(x+1)^5$ *b* $8(2x+3)^3$ *c* $3(5+x)^2$ *d* $24(3x-1)^7$

2 a $-4(1-x)^3$ *b* $-12(3-4x)^2$ *c* $-7(5-x)^6$ *d* $-20(2-4x)^4$

3 a $10x(x^2+1)^4$ *b* $12x^2(x^3-1)^3$ *c* $12x^3(x^4+8)^2$ *d* $-10x^4(1-x^5)$

4 a $(2x+3)^{-1/2}$ *b* $(3x-1)^{-2/3}$ *c* $-4(x-1)^{-5}$ *d* $6(2-3x)^{-3}$

5 a $3(x^2+3x)^2(2x+3)$ *b* $2(x^2+x+5)(2x+1)$ *c* $10(x^3-x)^9(3x^2-1)$

   d $(x+1)(x^2+2x+4)^{-1/2}$ *6 a* $(2x-5)^{-1/2}$ *b* $-2(1-4x)^{-1/2}$

   c $(x^2+1)(x^3+3x)^{-2/3}$ *d* $-\frac{1}{2}x(2-x^2)^{-3/4}$

7 a $-\dfrac{1}{(x+1)^2}$ *b* $\dfrac{5}{(1-x)^2}$ *c* $-\dfrac{4}{(2x+3)^3}$ *d* $-\dfrac{8}{(4x+1)^{3/2}}$

8 *a* $4x(x^2+2)$ *b* $2\left(x+1+\dfrac{1}{x}\right)\left(1-\dfrac{1}{x^2}\right)$ *c* $8\left(x^2+1+\dfrac{1}{x^2}\right)\left(x-\dfrac{1}{x^3}\right)$

9   31, maximum   *10* -2, minimum

Page 237   Exercise 5

1 *a* $3\cos 3x$ *b* $4\cos(4x+1)$ *c* $a\cos(ax+b)$ *d* $\tfrac{1}{2}\cos\tfrac{1}{2}x$

2 *a* $-2\sin 2x$ *b* $-3\sin(3x+4)$ *c* $-p\sin(px+q)$ *d* $-\tfrac{1}{2}\sin\tfrac{1}{2}x$

3 *a* $2\sin x\cos x$ *b* $3\sin^2 x\cos x$ *c* $-2\cos x\sin x$ *d* $-4\cos^3 x\sin x$

4 *a* $2(1+\sin x)\cos x$ *b* $3(1-\cos x)^2\sin x$ *c* $-2\cos x/\sin^2 x$ *d* 0

5 *a* F *b* T *c* T *d* F

6    $f'(x)=1-\sin\tfrac{1}{2}x\geqslant 0;\ \pi$ 7 *a* $\tfrac{1}{2}\pi,\tfrac{3}{2}\pi$ *b* $\tfrac{1}{3}\pi,\pi,\tfrac{5}{3}\pi$

8    $(0,0),(\tfrac{1}{2}\pi,0),(\pi,0);(\tfrac{1}{4}\pi,1),(\tfrac{3}{4}\pi,-1)$

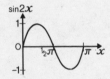 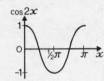

9    $(\tfrac{1}{4}\pi,0),(\tfrac{3}{4}\pi,0);(0,1),(\tfrac{1}{2}\pi,-1),(\pi,1)$

10*a* $\dfrac{\pi}{180}\cos x°$ *b* $-\dfrac{\pi}{180}\sin x°$

Page 239   Exercise 6

1 *a* $\tfrac{1}{5}(x+3)^5+C$ *b* $\tfrac{1}{14}(2x+1)^7+C$ *c* $\tfrac{1}{5a}(ax+b)^5+C$ *d* $-\tfrac{1}{8}(1-2x)^4+C$

2 *a* $-(x+4)^{-1}+C$ *b* $-\tfrac{1}{9}(3x+2)^{-3}+C$ *c* $\tfrac{1}{3}(2x-1)^{3/2}+C$ *d* $-\tfrac{1}{5}(1-2x)^{5/2}+C$

3 *a* $-(x+1)^{-1}+C$ *b* $-(x-3)^{-2}+C$ *c* $(2x+5)^{1/2}+C$ *d* $3(x-1)^{1/3}+C$

4 *a* 10 *b* 4 *c* $5\tfrac{1}{3}$ 5 *a* $\tfrac{1}{12}$ *b* 2 *c* $136\tfrac{2}{5}$

Page 239   Exercise 7

1 *a* $\tfrac{1}{2}\sin 2x+C$ *b* $\tfrac{1}{4}\sin 4x+C$ *c* $\sin(x+5)+C$,   *d* $\tfrac{1}{3}\sin(3x+2)+C$

2 *a* $-\tfrac{1}{3}\cos 3x+C$ *b* $\tfrac{1}{5}\cos 5x+C$ *c* $-2\cos\tfrac{1}{2}x+C$ *d* $\cos(2-x)+C$

3 *a* $\tfrac{1}{3}$ *b* $\tfrac{1}{2}$ *c* $\tfrac{1}{3}$ 4 *a* -1 *b* $\tfrac{7}{24}$

5   0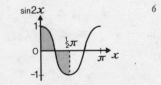

6   1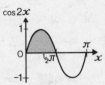

7 a (1) $\frac{1}{2}\sin 2x$ (2) $\frac{1}{2}(1+\cos 2x)$ (3) $\frac{1}{2}(1-\cos 2x)$ (4) $\cos 2x$

b (1) $-\frac{1}{4}\cos 2x + C$ (2) $\frac{1}{2}x + \frac{1}{4}\sin 2x + C$ (3) $\frac{1}{2}x - \frac{1}{4}\sin 2x + C$ (4) $\frac{1}{2}\sin 2x + C$

8 $\frac{1}{2}(\cos 4x + \cos 2x);\ \frac{1}{8}\sin 4x + \frac{1}{4}\sin 2x + C$

9 a $\frac{1}{2}(\sin 6x + \sin 2x);\ -\frac{1}{12}\cos 6x - \frac{1}{4}\cos 2x + C$

b $\frac{1}{2}(\cos 3x + \cos x);\ \frac{1}{6}\sin 3x + \frac{1}{2}\sin x + C$

c $\frac{1}{2}(\cos 4x - \cos 6x);\ \frac{1}{8}\sin 4x - \frac{1}{12}\sin 6x + C$

10 $\frac{1}{2}\pi^2$ 　　11 $\frac{1}{12}\pi^2$ 　　12c $y = 2\cos nx + \dfrac{1}{n}\sin nx$

Calculus – Answers to Revision Exercise

Page 242 　　*Revision Exercise 1*

1 a 0 　b $\frac{1}{2}\sqrt{3}$ 　c $-\frac{1}{2}$ 　d $\frac{1}{2}\sqrt{3}$ 　e 0

2 a $10\cos x$ 　b $-8\sin x$ 　c $2-3\cos x$ 　d $2x + \sin x$

3 a $-3\sin x - 4\cos x$ 　b $2\cos x - 2$ 　c $-\cos x + \sin x$

4 a (1) 1 　(2) $1-\frac{1}{\sqrt{2}}$ 　(3) 0 　(4) $1-\frac{1}{\sqrt{2}}$ 　(5) 1 　b $(\frac{1}{2}\pi, \frac{1}{2}\pi)$ 　c point of inflexion

5 a $(\frac{1}{4}\pi, 0),\ (\frac{5}{4}\pi, 0)$
b $(\frac{3}{4}\pi, \sqrt{2})$, maximum;
$(\frac{7}{4}\pi, -\sqrt{2})$, minimum.

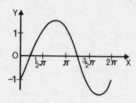

6 a $3\sin x + C$ 　b $C - 7\cos x$ 　c $\sin x + \cos x + C$ 　d $x + \frac{1}{2}x^2 + \sin x + C$

e $\frac{1}{3}x^3 - 2\sin x + C$ 　f $C - 6\cos x - 8\sin x$

7 a $\frac{3}{2}$ 　b $\sqrt{2} - 1$ 　c 2 　d $\frac{4}{9}\pi^2 - \frac{1}{2}\sqrt{3}$ 　e 7

8 $\frac{1}{2}\pi + 1$ 　　9 $5 + \frac{5}{2}\sqrt{3} - \frac{25}{24}\pi$

10a $9(3x+4)^2$ 　b $14(2x-1)^6$ 　c $6(1-2x)^{-4}$ 　d $6x(x^2+1)^2$

e $-54x(1-3x^2)^8$ 　f $(2x-1)^{-1/2}$ 　g $\frac{3}{2}x^2(x^3+1)^{-1/2}$ 　h $-8(x+1)(x^2+2x)^{-5}$

11a $2\cos(2x-1)$ 　b $-9\sin(3x+2)$ 　c $2x\cos(x^2)$ 　d $-(x+1)\sin(x^2+2x)$

e $4\sin^3 x \cos x$ 　f $-3\cos^2 x \sin x$ 　g $8\cos x(2\sin x + 1)^3$

h $-\sin x(2 - 3\cos x)^{-4/3}$

12a $\frac{1}{24}(4x+3)^6 + C$ 　b $\frac{1}{3}(2x+1)^{3/2} + C$ 　c $\frac{1}{10}(2-5x)^{-2} + C$ 　d $C - (3x+2)^{-2/3}$

13a $C - \frac{1}{4}\cos(4x+3)$ 　b $C - \frac{1}{2}\sin(1-2x)$ 　c $C - \frac{1}{3}\cos 3x + \frac{1}{4}\sin 4x + 5x$

14a $8\frac{2}{3}$ 　b 4 　c $\frac{2}{3}$ 　　15a $21\frac{1}{4}$ 　b $\frac{5461}{7}\pi$ 　　16 -2; maximum

18 $0\cdot 34$ radian, $\frac{1}{2}\pi$ radian; $31\cdot 7$s, $15\cdot 7$s

Calculus – Answers to Cumulative Revision Section

Page 248   Exercise 1

1 a 4   b 6x   c 2   d $-4/x^2$ e $2\cos 2x$ f $-3\sin 3x$

2   72 m/s   3 a $15x^2+8x+3$ b $4t^3-4t$ c $2u^2+3u$

d $6x^{1/2}-x^{-1/2}$ e $-1+\dfrac{1}{x^2}$ f $-\dfrac{2}{x^3}-\dfrac{3}{x^4}$ g $\dfrac{1}{2\sqrt{x}}+\dfrac{1}{2\sqrt{x^3}}$

h $-\dfrac{3}{x^2}+\dfrac{1}{x^3}$ i 1   4 a $6(3x+4)$ b $-24(5-4x)^5$

c $\frac{3}{2}(3x-1)^{-1/2}$ d $4x(6x^2-5)^{-2/3}$ e $3x^2+4x+1$ f 1

g $-3(3x-5)^{-2}$ h $-2(x+5)^{-3}$ i $(3-x)^{-3/2}$ j $-1/x^2$

k $1-2/x^2$ l $-\frac{1}{2}x^{-3/2}-x^{-2}$ 5 a $3\cos 3x$ b $p\cos(px+q)$

c $3\cos x+4\sin x$ d $m\cos mx-n\sin nx$ e $-2\cos x\sin x$

f $3\sin^2 x\cos x$ g $4\sin^3 x\cos x+4\cos^3 x\sin x$ h $-\dfrac{\cos x}{\sin^2 x}$ i $\dfrac{\sin x}{\cos^2 x}$

Page 249   Exercise 2

1   4, −2, 16   2 a $\frac{3}{4}$, −3   b ±1, ±2   3 $f'(x)=3(x-2)^2$, which
is never negative.   4 (−1, 9), (3, −23)   a $x<-1$ and $x>3$ b $-1<x<3$

5 a $13x+y=2$ b $y-\frac{1}{\sqrt{2}}=-\frac{1}{\sqrt{2}}(x-\frac{1}{4}\pi)$ c $4y-15x=-4$

d $y=2x+4$ 6 1, −5   7 7, −9   8 32, 5

9 a 2   b 1   10 (0, 0), $(\frac{1}{2}\pi, 1)$, $(\pi, 0)$, $(\frac{3}{2}\pi, 1)$, $(2\pi, 0)$

a $0<x<\frac{1}{2}\pi$, and $\pi<x<\frac{3}{2}\pi$

b $\frac{1}{2}\pi<x<\pi$, and $\frac{3}{2}\pi<x<2\pi$

11a 4, max; 0, min   b 4, max; 0, min   c 16 max; 0 min   d 1, max; −1, min

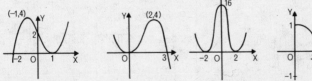

12 (3, 0), (1, −8)

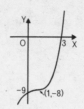

13a $f(-2) = -1, f(-1) = 3$

$f(0) = 1, f(1) = -1, f(2) = 3$

b (−1, 3), max; (1, −1), min.

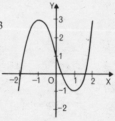

14a 0 m/s (at max. height); −20 m/s (i.e. downwards)   b (1) 0, 6   (2) 3

c 45 m   15a 6 cm, 30 cm   b 16 cm/s, 33 cm/s; 14 cm/s², 20 cm/s²

c $1\frac{3}{4}$ cm/s   16 length and breadth both $2\sqrt{3}$ cm   17 25 cm²

18   height $2\sqrt{3}$ cm, radius of base $2\sqrt{6}$ cm

Page 252   Exercise 3

1 a $2x^4 + C$ b $\frac{1}{3}x^3 - 5x + C$ c $2x^5 + 2x^2 + C$ d $-x^{-1} + C$

e $x^4 + 2x^{-2} + C$ f $\frac{1}{3}\sin 3x + C$ g $\sin x + \cos x + C$ h $-\frac{1}{4}\cos(4x - 3\pi) + C$

i $-\frac{1}{2}\sin(\pi - 2x) + C$ j $-\frac{1}{4}\cos 4x - \frac{1}{4}\sin 4x + C$ k $\frac{1}{2}x + \frac{1}{4}\sin 2x + C$

l $\frac{1}{2}x - \frac{1}{4}\sin 2x + C$

2 a $x^5 - x^3 + x + C$ b $t + t^2 + t^3 + t^4 + C$ c $\frac{1}{5}u^4 - \frac{1}{4}u^5 + C$

d $2x^{5/2} - 10x^{3/2} + C$ e $-\frac{1}{3}x^3 + \frac{5}{2}x^2 - 6x + C$ f $-\frac{3}{4}v^4 - v^3 + 2v^2 + 4v + C$

g $\frac{1}{2}x^2 - \frac{4}{3}x^{3/2} + x + C$ h $-u^{-1} + \frac{1}{2}u^{-2} + C$ i $\frac{2}{3}z^{3/2} + 2z^{1/2} + C$

3 a $2x^2 - x + 2$ b $\frac{1}{3}x^3 - x^{-1} + 1\frac{2}{3}$

4 a   b   c   d

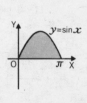

5   a $\frac{1}{10}(2x+3)^5+C$ b $\frac{1}{21}(3u-5)^7+C$ c $-\frac{1}{3}(5-2t)^{3/2}+C$

   d $2(4+x)^{1/2}+C$ e $\frac{1}{2}(1-z)^{-2}+C$ f $\frac{1}{3}\sin(3x-\frac{1}{4}\pi)+C$

   g $\frac{1}{4}x^4+\cos x+C$ h $\frac{1}{2}\sin 2x-\cos x+C$ i $\sin x+\frac{1}{3}\cos 3x+C$

   j $\frac{1}{12}\sin 6x+\frac{1}{4}\sin 2x+C$ k $-\frac{1}{8}\cos 4x-\frac{1}{4}\cos 2x+C$ l $\frac{1}{8}\sin 4x-\frac{1}{16}\sin 8x+C$

6   a $1\frac{5}{6}$ b 1 c $4\frac{2}{3}$ d 2 e 2 f $\sqrt{2}$ g $\frac{1}{8}\pi+\frac{1}{4}$ h $\frac{1}{4}$ i $-\frac{2}{3}$

Page 253   Exercise 4

1    $10\frac{2}{3}$ 2 $6\frac{3}{4}$ 3 $3\frac{1}{12}$ 4 $6\frac{3}{4}$ 5   36   6   1   7   2

8    $y=\frac{1}{3}(2x+1)^{3/2}+1$ 9 $1\,\text{m/s};\frac{2}{3}\,\text{m}$ 10 2π 11 $\frac{1}{2}\pi^2$

12    $\frac{1}{3}\pi$ 13 $10\frac{2}{3};51\frac{1}{3}\pi$ 14 $\frac{1}{24}\pi^2$